First International Seminar on Physical Infrastructure (1st ISPI)

Journal of Physics: Conference Series Volume 2153

San Jose de Cucuta, Colombia
13 – 15 October 2021

ISBN: 978-1-7138-5124-0
ISSN: 1742-6588

Printed from e-media with permission by:

Curran Associates, Inc.
57 Morehouse Lane
Red Hook, NY 12571

Some format issues inherent in the e-media version may also appear in this print version.

Institute of Physics
Dirac House, Temple Back
Bristol BS1 6BE UK

Phone: 44 1 17 929 7481
Fax: 44 1 17 920 0979

techtracking@iop.org

Additional copies of this publication are available from:

Curran Associates, Inc.
57 Morehouse Lane
Red Hook, NY 12571 USA
Phone: 845-758-0400
Fax: 845-758-2633
Email: curran@proceedings.com
Web: www.proceedings.com

TABLE OF CONTENTS

First International Seminar on Physical Infrastructure (1st ISPI)

Y W Yung-Vargas[1], A Rodríguez-Lizcano[1], L F Ortiz-Vasquez[2], J P Rojas-Suárez[1], and E D V-Niño[3]
[1] Universidad Francisco de Paula Santander, San José de Cúcuta, Colombia
[2] Corporación Universitaria Minuto de Dios - UNIMINUTO, Bello, Colombia
[3] Foundation of Researchers in Science and Technology of Materials, Bucaramanga, Colombia

E-mail: info@foristom.org

The First International Seminar on Physical Infrastructure (1st ISPI) was an academic and scientific event organized by the Road Infrastructure Research Group and the Road Infrastructure Research Seedbed of the Universidad Francisco de Paula Santander, San José de Cúcuta, Colombia; the 1st ISPI was held from October 13 to 15, 2021.

The purpose of the 1st ISPI was to allow the academic and scientific community, researchers, professors, professionals, and students, to share results and research proposals in the areas of Engineering, Mathematics, and Physics; during the 1st ISPI is presented 29 research work, of which 6 were plenary lectures from Spain, Brazil, Venezuela, and Colombia, and 23 talks, on different topics that allowed to the participants the to learn about the current state of the Physical Infrastructure, the Mathematical Modeling, the Physical and Numerical Modeling, the Physicochemical Phenomena in Engineering, the Advances in Materials Science and Engineering, the Simulation of Physical-Mathematical Problems in Engineering, and the Teaching and Research Techniques in Science and Engineering. The website of the conference is available at https://foristom.org/1ispi.

On behalf of the organizing committee of the 1st ISPI, we are extremely thankful to all authors and participants for providing their valuable contributions to this Proceedings volume as well as the reviewers for their constructive recommendations and criticism aiding to improve the presented articles. Likewise, we like to thank the Universidad Francisco de Paula Santander, and the Foundation of Researchers in Materials Science and Technology, for all the support technical and logistical received.

Likewise, the organizing committee 1st ISPI thanks all who attended this edition of the event, and we are very proud to have carried out this event successfully that allowing the generation and effective transfer of new knowledge regarding the topics addressed during the event; we will be waiting for you at the 2nd ISPI in 2022.

Finally, the editor hopes that those interested in the area of physical infrastructure can enjoy this reading, of the volume of the Journal of Physics: Conference Series (JPCS), which reflects a wide variety of current issues.

"Education and expression freedom for anyone".

Organizing Committee

Yee Wan Yung Vargas (Chairman)

Adriana Rodriguez Lizcano (Co-Chairman)
Jhan Piero Rojas Suárez

National Scientific Committee

- Claudia Helena Muñoz Hoyos
  Universidad de Antioquia, Medellín
- Margarita Rosa Gómez Betancur
  Universidad Nacional de Colombia, Medellín
- Vanessa Senior Arrieta
  Universidad Cooperativa de Colombia, Medellín
- Hugo Alexander Rondón Quintana
  Universidad Distrital Francisco José de Caldas, Bogotá
- Carlos Alberto Peña Soto
  Universidad Francisco de Paula Santander, San José de Cúcuta
- Adriana Rodríguez Lizcano
  Universidad Francisco de Paula Santander, San José de Cúcuta
- Fernando Jaimes Tarazona
  Universidad Francisco de Paula Santander, San José de Cúcuta
- Andrea Jovanna Cacique Arias
  Universidad Francisco de Paula Santander, San José de Cúcuta
- Jorge Arias
  Universidad Francisco de Paula Santander, San José de Cúcuta
- Diana Patricia López Gómez
  Universidad Francisco de Paula Santander, San José de Cúcuta
- Jorge Iván Arias Flórez
  Universidad Francisco de Paula Santander, San José de Cúcuta
- Yee Wan Yung Vargas
  Universidad Francisco de Paula Santander, San José de Cúcuta
- Dr. Ely Dannier V. Niño
  Foundation of Researchers in Science and Technology of Materials

International Scientific Committee

- Francisco de Paula Montero Chacón
  Universidad Loyola Andalucia, Spain
- Griselda Lopez Maldonado
  Universidad Politécnica de Madrid, Spain
- Marcio Muniz de Farias
  Universidad de Brasilia, Brazil
- Luis Ángel Moreno Anselmi
  Universidad de Brasilia, Brazil
- Linda Elcida Gil Lozada
  Universidad Nacional Experimental Politécnica Antonio José de Sucre, Venezuela
- José Rubén Calderas Volcanes
  Universidad de Los Andes de Mérida, Venezuela
- Dr. (c) Anderson Sandoval Amador
  Universidad Nacional del Centro de la Provincia de Buenos Aires, Argentina

Sponsor

- ✓ Universidad Francisco de Paula Santander

Partners

- ✓ Semillero de Investigación en Infraestructura Vial, Universidad Francisco de Paula Santander
- ✓ Grupo de Investigación en Infraestructura Vial, Universidad Francisco de Paula Santander
- ✓ Foundation of Researcher in Science and Technology of Materials
- ✓ IOP Publishing

Journal of Physics: Conference Series **2153** (2022) 011001 doi:10.1088/1742-6596/2153/1/011001

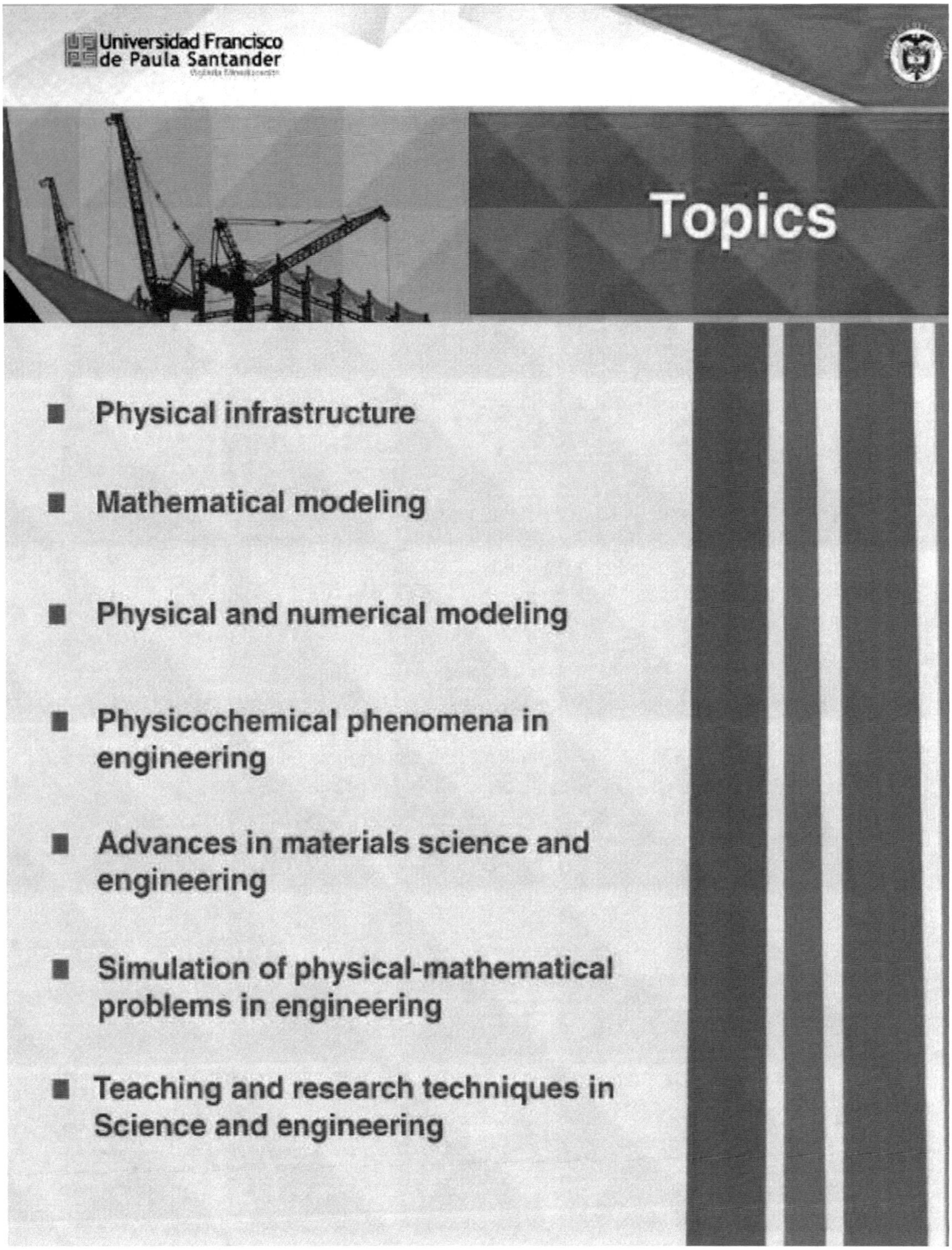

Universidad Francisco de Paula Santander
International Seminar on Physical Infrastructure
Sponsors
Universidad Francisco de Paula Santander
Vigilada Mineducación
Partners
GIN FRA VIAL
Grupo de Investigación en Infraestructura Vial
SinfraVIAL
Semillero de Investigación en Infraestructura Vial
Foristom Foundation
of Researchers in Science and Technology of Materials
IOP science

Journal of Physics: Conference Series **2153** (2022) 011002 doi:10.1088/1742-6596/2153/1/011002

Peer review declaration

All conference organisers/editors are required to declare details about their peer review. Therefore, please provide the following information:

- **Type of peer review: Single-blind**

- **Conference submission management system:**
  By email received from the organizing and editorial committee of the conference. The correspondence authors make the submission by email.

- **Number of submissions received:**
  22

- **Number of submissions sent for review:**
  22

- **Number of submissions accepted:**
  20

- **Acceptance Rate (Number of Submissions Accepted / Number of Submissions Received X 100):**
  91.91%

- **Average number of reviews per paper:**
  2 reviews per paper

- **Total number of reviewers involved:**
  11

- **Any additional info on review process:**
  All papers are plagiarism checked by a plagiarism software

- **Contact person for queries:**
  Ely Dannier V. Niño
  Foundation of Researchers in Science and Technology of Materials (FORISTOM)

Assessment of the asphalt mixtures properties subjected to a flexural strength

J C Ruge[1], **H A Rondon-Quintana**[2], **and J G Bastidas-Martínez**[3]
[1] Universidad Militar Nueva Granada, Bogotá, Colombia
[2] Universidad Distrital Francisco José de Caldas, Bogotá, Colombia
[3] Universidad Piloto, Bogotá, Colombia

E-mail: juan.ruge@unimilitar.edu.co

Abstract. Fatigue cracking by loading is one of the main mechanisms of damage to asphalt mixtures in service. Several studies worldwide have been conducted to try to understand the response that hot-mix asphalt undergo under this mechanism of damage. Despite the above, the fatigue phenomenon in asphalt mixtures is still not fully understood. The current research hypothesizes that the response under repeated loading of asphalt mixtures in fatigue tests can be more clearly understood through the one obtained under monotonic loading. For this reason, this study presents the results of the first phase of the research in which beams of asphalt mixtures were subjected to flexion using monotonic loads. The above, to correlate the evaluated properties with those obtained in a second phase where the response of the beams under repeated load (fatigue) will be measured. Beams made of two hot-mix asphalt mixes, two asphalt contents, and two different thicknesses were subjected to flexural strength tests. From the tests, the modulus of rupture, the maximum monotonic load that supports the beams in the failure state, the displacement in the failure state, and the relation between load and displacement were obtained. As a general conclusion of the study, it was obtained that the response experienced by the beams subjected to monotonic load has a broad correlation with the reported in the reference literature.

1. Introduction

One of the primary damage mechanisms of asphalt mixtures is fatigue cracking [1-3]. In Asphalt pavements, fatigue cracking due to loads occurs when high accumulated tensile strain values are generated in the lower area of the asphalt layer due to the loads imposed by the traffic (repeated loads). It is considered that around the 1940s, the exponential increase in vehicles began in the world [4]. Practically, since that date, it has been reported that the asphalt layers begin to fail at minimal strains (deflections), close to 0.5 mm [5].

The continuous action of traffic typically exceeds the flexural strength in these materials. This effect is revealed with the appearance of cracks at the end of the lifespan [6]. The continuous movement of vehicles causes the asphalt layer to flex. This repetition of loading makes the asphalt mix lose stiffness and causes the appearance of tensile plastic strains, which simultaneously leads to the formation of microcracks. Due to the continuity of the load repetitions and the decrease in stiffness, the microcrack coalescence occurs, leading to the formation of a visible fissure (macrocrack) at ground level. Once the crack appears superficially, the life of the pavement decreases exponentially because water and air penetrate more quickly [7].

Despite the large number of research carried out on the subject, the phenomenon of fatigue in asphalt mixtures is still not fully understood, due to the multiple variables that affect the response that these materials undergo under repeated load. Some of them are the magnitude and speed of the load, the type of aggregates and asphalt, temperature, stiffness, and dimensions of the mixture, effects associated with aging and moisture damage, among others [8-12]. The test methods used to measure the fatigue resistance of asphalt mixtures can be grouped into the following categories [13]: Simple bending tests, bending tests with supports, direct axial tests, diametric tests, triaxial tests, and track tests.

In the previous tests (under controlled stress or strain), the number of load cycles that the material can resist until failure can be estimated [14,15]. Among the simple bending tests, one of the most used to evaluate the fatigue life of asphalt mixtures under repeated load is the bending beam test [3] (generally with load in the center of the beam or thirds). The present study hypothesizes that these tests of fatigue resistance on beams can be understood more clearly through the response of asphalt mixtures subjected to monotonic loads. Therefore, an experimental phase was designed to correlate the response obtained from flexural beam tests under monotonic load with the one obtained from fatigue resistance tests under repeated load.

For that purpose, the study was divided into two phases; this document presents the results of the first phase of the study, in which flexural resistance tests were carried out under monotonic load on beams made with two hot mix asphalt (HMA), two asphalt contents, and two different dimensions (varying the beam height). The current study is the first point of analysis for long-term research, which seeks to analyze the fatigue failure in beams of asphalt mixtures subjected to repetitive loads. Similar research to the one described in the present study were not found in the literature consulted. In this sense, the present document contributes to the state-of-the-art.

2. Methodology and materials

In Table 1 and Table 2, the asphalt AC 60-70 (0.1 mm) and aggregate properties used in the fabrication of beams are shown. These materials meet the established requirements by [16] to be used in manufacturing mixtures HMA.

Table 1. Physical characteristics of asphalt AC 60-70.

Characteristics	Specification	Measure	AC 60-70		
			Minimum	Maximum	Result
Penetration	[17]	0.1mm	50	70.0	67.00
Penetration index	[18]	-	-1.2	0.6	-0.59
Softening point	[19]	°C	48	54.0	49.60
Specific gravity	[20]	-	-	-	1.021
Viscosity Absolute 60°C	[21]	P	1500	-	2172
Flashpoint	[22]	°C	230	-	302
Ductility	[23]	cm	100	-	135
Loss of mass	[24]	%	-	0.8	0.40
Penetration, in% with refence to the original penetration	[17]	%	50	-	57.50
Increase of softening point	[18]	°C	-	9.0	7.80

Table 2. Physical characteristics of aggregate.

Characteristics	Standard	Measure	Value	Specification
Sand equivalent	[25]	%	66.0	min 55.0
Abrasion in Los Angeles	[26]	%	21.0	max 50.0
Form index	[27]	-	0.7	min 0.5
Soundness of aggregates by use of magnesium sulfate	[28]	%	7.0	max 12.0
Bulk specific gravity coarse	[29]	-	2.6	-
Bulk specific gravity saturated- surfaced –dry coarse	[29]	-	2.6	-
Absorption coarse	[29]	%	0.8	-
Specific gravity of fine aggregates	[30]	-	2.6	-

2.1. Marshall test
Marshall tests [31] were carried out to obtain the asphalt percentages to use in the manufacturing of the beams. The particle size distribution (Figure 1) recommended by [16] was used to manufacture the Marshall samples HMA-19 and HMA-25. In order to measure the volumetric composition of the mixtures (air void content (Va), voids in mineral aggregate (VMA), and the voids filled with asphalt (VFA)), and the strength parameters under monotonic load (stability (S), flow (F), and stability-flow ratio (S/F)) (see Table 3).

Five samples of 1200 g were manufactured for each percentage of asphalt of 4.5%, 5.0%, 5.5%, and 6.0% (concerning the total mass). Each sample was compacted by applying 75 blows per face. Mixing and compaction temperatures were 150 °C (asphalt viscosity = 170 cP) and 140 °C (asphalt viscosity = 280 cP), respectively. The results of the Marshall test are presented in Table 1. The optimal asphalt contents (OAC) of the HMA-19 and HMA-25 mixtures are 5.3% and 5.0%, respectively. Therefore, the decision to manufacture the beams with asphalt contents of 5.0% and 5.5% were made in both HMAs.

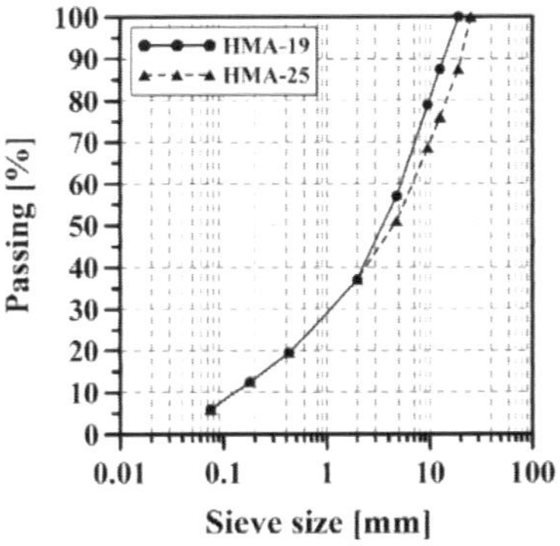

Figure 1. Particle size distribution curves of HMA mixtures.

Table 3. Marshall test.

Asphalt (%)	S (KN)	F (mm)	S/F (KN/mm)	Bulk density (g/cm^3)	Va (%)	VMA (%)	VFA (%)
HMA-19							
4.5	11.3	4.06	2.8	2.305	6.0	16.30	63.0
5.0	12.5	3.98	3.2	2.321	4.7	16.20	70.9
5.5	12.8	3.94	3.3	2.329	3.7	16.30	77.6
6.0	11.5	4.15	2.8	2.323	3.2	17.00	81.2
HMA-25							
4.5	11.9	3.76	3.2	2.285	4.6	14.77	68.8
5.0	13.7	3.80	3.6	2.284	4.0	15.25	74.0
5.5	13.5	3.84	3.5	2.288	3.1	15.53	80.1
6.0	12.9	4.14	3.1	2.325	0.9	14.63	94.2

2.2. Beams manufacturing and flexural strength test
To carry out the flexural strength tests, 32 beams were manufactured: 4 samples for each asphalt content (5.0% and 5.5%), beam height d (5 cm and 10 cm), and type of mixture (HMA-19 and HMA-25). The length and width (b) of each beam were 42 cm and 15 cm, respectively. The length between supports (L) was 38 cm. The particle size distribution (Figure 1) and the mixing and compaction temperatures to manufacture the beams were the same as the Marshall test (150 ° C and 140 °C, respectively). The beams were compacted by applying a monotonic load on their surface until achieving Va, obtained from the Marshall test. The above process was performed by trial and error on beams that were not failed (*i.e.*, beams manufactured only to achieve the standard dimensions and Va). An example of a fabricated beam is presented in Figure 2(a).

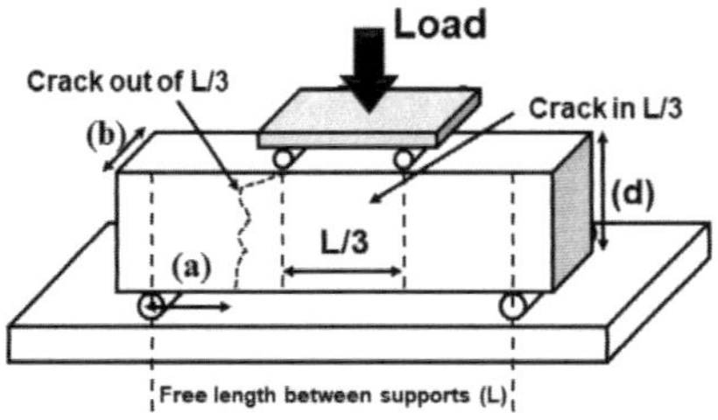

(a) (b)

Figure 2. (a) Example of manufactures beam; (b) test scheme.

The flexural strength tests on the beams were carried out following the guidelines established by [32]. Each beam was subjected to bending, applying a monotonic load (P) until failure; beam height (d) (5 cm and 10 cm), and type of mixture (HMA-19 and HMA-25). The length and width (b) of each beam were 42 cm and 15 cm, respectively. The length between supports (L) was 38 cm (see Figure 2(b)). According to [32], the displacement in a failure state (D) in mm, the distance between fracture line and closest support (a), were measured in the central area of each beam. With the results obtained from the tests, load-displacement ratio (P/D) (which can be understood as an indirect parameter of flexural stiffness under monotonic load) and the modulus of rupture (MR) were calculated using Equation (1).

$$MR = \frac{P \cdot L}{b \cdot d^2}. \quad (1)$$

In preliminary calibration tests, it was observed that the equipment is susceptible to loading speed. For very high speed, the record of maximum load P could occur suddenly during the test. For this reason, the speed was decreased to a value of 1.27 mm/min. Figure 2(b) shows the setup of the test.

3. Results and discussions

The average values (Av) with their respective standard deviation (SD) of the measurements and parameters obtained from the beams tested are shown in Table 2. It is observed that the MR tends to be higher in the HMA-25 mixtures. The above is perhaps due to the higher S/F ratio these mixtures undergo (greater stiffness and resistance under monotonic load) than HMA-19. This greater stiffness in tension as a product of the greater stiffness of the asphalt mix is coherent with other studies (*e.g.* [33,34]). Regardless of the above, the trend of the P/D relation is not clear. HMA-25 beams with a thickness of 10 cm undergo a higher P/D than HMA-19. However, when the thickness was 5 cm, the behavior was contrary. These results could help understand the response of asphalt mixtures in fatigue tests since the literature recommends tests with thin asphalt layers (≤5 cm) to carry out tests under controlled strain, while for thick asphalt layers, tests under controlled stress [35,36]. In both types of tests, the fatigue resistance of asphalt mixes is contrary. In fatigue tests under controlled strain, the mixtures that undergo better fatigue resistance are the minor stiffness. Nevertheless, under controlled stress, those with the increased stiffness [37,38]. On the other hand, it is evident that as the asphalt content increases, the D in the beams increases. Because when presenting a higher asphalt content, the mixtures tend to be more ductile. Also, HMA-25 mixes show higher D in the failure state as a product of the lowest Va and highest VFA.

4. Conclusions

This article shows the results of the first phase of a study to correlate the response that hot mix asphalt beams undergo in flexural strength tests under monotonic load. The above, to correlate this response with the one obtained under repeated load in fatigue tests in the future. Based on the results obtained, it is concluded that the beams manufactured with hot mix asphalt-25 undergo higher modulus of rupture compared to hot mix asphalt-19. The previous is perhaps due to the increased stiffness under monotonic load (S/F ratio) of the hot mix asphalt-25. This greater stiffness allowed the beams with hot mix asphalt-

25 and thickness of 10 cm to develop greater load-displacement ratio than those made with hot mix asphalt-19. A contrary behavior was experienced by the beams when the thickness was 5 cm, explaining the fatigue resistance in thin and thick asphalt layers, in which the strength is contrary (in thin asphalt layers, the mixtures experience less resistance to fatigue when their stiffness increases, while in thick the opposite occurs). Also, hot mix asphalt-25 mixes allow significant strain before failure, perhaps due to their lower air void content, and higher voids in mineral aggregate than hot mix asphalt-19. The higher the asphalt content, the more significant displacements are allowed in the failure, which is associated with increments in the mixture's ductility.

References

[1] Masad E, Castelo Branco V, Little D N, Lytton R 2008 A unified method for the analysis of controlled-strain and controlled-stress fatigue testing *International Journal of Pavement Engineering* **9(4)** 233
[2] Nejad F M, Aflaki E, Mohammadi M A 2010 Fatigue behavior of SMA and HMA mixtures *Construction and Building Materials* **24(7)** 1158
[3] Sudarsanan N, Arulrajah A, Karpurapu R, Amrithalingam V 2020 Fatigue performance of geosynthetic-reinforced asphalt concrete beams *Journal of Materials in Civil Engineering* **32(8)** 04020206
[4] Li N 2013 *Asphalt Mixture Fatigue Testing. Influence of Test Type and Specimen Size* (Delft: Delft University of Technology)
[5] Porter O J 1943 *Foundations for Flexible Pavements* (Washington: Proceedings Highway Research Board)
[6] Nijboer L W, Van der Poel C 1953 A study of vibration phenomena in asphaltic road constructions *Asphalt Paving Technology* **22** 197
[7] Rondón Quintana H A, Reyes F A 2015 *Pavimentos: Materiales, Construcción y Diseño* (Bogotá: ECOE)
[8] Moreno F, Rubio M C 2013 Effect of aggregate nature on the fatigue-cracking behavior of asphalt mixes *Materials and Design* **47** 61
[9] Moon K H, Falchetto A C, Marasteanu M O 2014 Investigation of limiting criteria for low temperature cracking of asphalt mixture *KSCE Journal of Civil Engineering* **18(1)** 172
[10] Meena P K, Saha G, Biligiri K P 2016 Estimation of fatigue life using resilient moduli of asphalt mixtures *Journal of Testing and Evaluation* **44(1)** 424
[11] Alinezhad M, Sahaf A 2019 Investigation of the fatigue characteristics of warm stone matrix asphalt (WSMA) containing electric arc furnace (EAF) steel slag as coarse aggregate and Sasobit as warm mix additive *Case Studies in Construction Materials* **11** e00265
[12] Ziari H, Amini A, Goli A 2020 The effect of different aging conditions and strain levels on relationship between fatigue life of asphalt binders and mixtures *Construction and Building Materials* **244** 118345
[13] Tangella S R, Craus J, Deacon J A, Monismith C L 1990 *Summary Report on Fatigue Response of Asphalt Mixtures* (Berkeley: University of California)
[14] Saha G, Biligiri K P 2015 Fracture damage evaluation of asphalt mixtures using semi-circular bending test based on fracture energy approach *Engineering Fracture Mechanics* **142** 154
[15] Zheng M, Li P, Yang J, Li H, Qiu Y, Zhang Z 2017 Fatigue life prediction of high modulus asphalt concrete based on the local stress-strain method *Applied Sciences* **7(3)** 305
[16] Instituto Nacional de Vías (INVIAS) 2013 *Mezclas Asfálticas en Caliente de Gradación Continua (Concreto Asfáltico), Artículo 450-13* (Colombia: Instituto Nacional de Vías)
[17] American Society for Testing and Materials (ASTM) 2020 *Standard Test Method for Penetration of Bituminous Materials, ASTM D5M-20* (United States of America: American Society for Testing and Materials)
[18] Centro de Estudios y Experimentación de Obras Públicas (CEDEX) 1999 *Indice de Penetración de los Betunes Asfálticos, NLT-181/99* (España: Centro de Estudios y Experimentación de Obras Públicas)
[19] American Society for Testing and Materials (ASTM) 2000 *Standard Test Method for Softening Point of Bitumen (Ring-and-Ball Apparatus), ASTM D36-95* (United States of America: American Society for Testing and Materials)
[20] American Association of State Highway and Transpotation Officials (AASHTO) 2009 *Standard Method of Test for Specific Gravity of Semi-Solid Asphalt Materials, AASHTO T 228-09* (United States of America: American Association of State Highway and Transpotation Officials)
[21] American Association of State Highway and Transpotation Officials (AASHTO) 2013 *Standard Method of Test for Viscosity Determination of Asphalt Binder Using Rotational Viscometer, AASHTO T 316-13* (United States of America: American Association of State Highway and Transpotation Officials)

[22] American Society for Testing and Materials (ASTM) 2019 *Standard Test Method for Flash Point of Cutback Asphalt with Tag Open-Cup Apparatus, ASTM D3143M-19* (United States of America: American Society for Testing and Materials)
[23] American Society for Testing and Materials (ASTM) 2017 *Standard Test Method for Ductility of Asphalt Materials, ASTM D113-17* (United States of America: American Society for Testing and Materials)
[24] American Society for Testing and Materials (ASTM) 2019 *Standard Test Method for Effect of Heat and Air on a Moving Film of Asphalt (Rolling Thin-Film Oven Test), ASTM D2872-19* (United States of America: American Society for Testing and Materials)
[25] American Society for Testing and Materials (ASTM) 2009 *Standard Test Method for Sand Equivalent Value of Soils and Fine Aggregate, ASTM D2419-09* (United States of America: American Society for Testing and Materials)
[26] American Society for Testing and Materials (ASTM) 2006 *Standard Test Method for Resistance to Degradation of Small-Size Coarse Aggregate by Abrasion and Impact in the Los Angeles Machine, ASTM C131-06* (United States of America: American Society for Testing and Materials)
[27] Departamento Nacional de Estradas de Rodagem (DNER) 1994 *Agregado - Determinação do Índice de Forma, DNER-ME 086/94* (Brazil: Departamento Nacional de Estradas de Rodagem)
[28] American Society for Testing and Materials (ASTM) 2005 *Standard Test Method for Soundness of Aggregates by Use of Sodium Sulfate or Magnesium Sulfate, ASTM C88-05* (United States of America: American Society for Testing and Materials)
[29] American Society for Testing and Materials (ASTM) 2007 *Standard Test Method for Density, Relative Density (Specific Gravity), and Absorption of Coarse Aggregate, ASTM C127-07* (United States of America: American Society for Testing and Materials)
[30] American Society for Testing and Materials (ASTM) 2007 *Standard Test Method for Density, Relative Density (Specific Gravity), and Absorption of Fine Aggregate, ASTM C128-07a* (United States of America: American Society for Testing and Materials)
[31] American Association of State Highway and Transpotation Officials (AASHTO) 2015 *Standard Method of Test for Resistance to Plastic Flow of Asphalt Mixtures Using Marshall Apparatus, AASHTO T 245-15* (United States of America: American Association of State Highway and Transpotation Officials)
[32] American Association of State Highway and Transpotation Officials (AASHTO) 2018 *Standard Method of Test for Flexural Strength of Concrete (Using Simple Beam with Third-Point Loagind), AASHTO T 97-18* (United States of America: American Association of State Highway and Transpotation Officials)
[33] Molenaar A A A, Li N 2014 Prediction of compressive and tensile strength of asphalt concrete *International Journal of Pavement Research & Technology* **7(5)** 324
[34] Teltayev B, Radovskiy B 2018 Predicting thermal cracking of asphalt pavements from bitumen and mix properties *Road Materials and Pavement Design* **19(8)** 1832
[35] Elseifi M A, Flintsch G W, Al-Qadi I L 2003 Quantitative effect of elastomeric modification on binder performance at intermediate and high temperatures *Journal of Materials in Civil Engineering* **15(1)** 32
[36] Muniz de Farias M, Quiñonez F, Rondón H A 2019 Behavior of a hot-mix asphalt made with recycled concrete aggregate and crumb rubber *Canadian Journal of Civil Engineering* **46(6)** 544
[37] Rondón-Quintana H A, Ruge-Cárdenas J C, Bastidas-Martínez J G, Velandia-Castelblanco M Y, Muniz de Farias M 2020 Use of thermally treated bentonite as filler in hot mix asphalt *Journal of Materials in Civil Engineering* **32(5)** 04020070
[38] Bastidas-Martínez J G, Rodríguez-Joaquín N, Torres-Daza H D, Rondón-Quintana H A, Ruge-Cárdenas J C 2020 Behavior of a draining mixture composed by recycled concrete aggregates and rubberized asphalt concrete *Respuestas* **25(1)** 96

Numerical simulation of box culverts subjected to different physical actions and design regulations

M V Afanador Arias[1], J S Pacheco-Posada[1], and N Afanador García[2]

[1] Programa de Ingeniería Civil, Universidad Francisco de Paula Santander, Seccional Ocaña, Colombia

[2] Civil Engineering Research Group, Universidad Francisco de Paula Santander, Seccional Ocaña, Colombia.

E-mail: mvafanadora@ufpso.edu.co, nafanadorg@ufpso.edu.co

Abstract. This research work was intended to provide a broad and sufficient solution to areas of difficult vehicular traffic, where during the rainy season some watersheds present a non-negligible water flow. The research carried out the structural design of different geometries of box culverts, to provide a solution to vehicular continuity in areas affected by water flows that interrupt pedestrian and vehicle traffic, endangering human lives. The structural designs followed the standards established by the American Association of State Highway and Transportation Officials and the Colombian bridge code. The designs were carried out using commercial software SAP2000 for analysis and design through Visual Basic programming, allowing a wide range of geometries and boundary conditions to be established. The structural designs indicated a significant difference in quantities of work in the construction of the box culverts for different geometries, indicating higher volumes of concrete for the designs made using the American Association of State Highway and Transportation Officials code compared to Colombian bridge code. The Colombian bridge code has higher steel requirements for fills less than 3.00 m, while for greater heights the American Association of State Highway and Transportation Officials code has the highest steel requirements.

1. Introduction

Box culverts are low structures composed of two horizontal and two vertical slabs built monolithically. They are ideal for the crossing of railways, roads or overpasses with drainage areas or water channels, and are also used in areas where the bearing capacity of the soil is low; the box culverts is an ideal bridge structure. Culverts are reinforced concrete structures with square or rectangular openings, with spans of up to 4.00 m, clear heights generally not exceeding 3.00 m and embankment heights of up to 6.00 m [1].

The box culvert is only a name given to it because of its shape, although it can be found less frequently in various types of shapes. These structures are economical, have great rigidity due to their monolithic action and do not require isolated foundations, since the bottom slab rests directly on the ground. For small discharges, single cell or box culverts are used; for large discharges, multi-cell or multi-bin culverts are usually used [2]. Box culverts are designed so that the top slab supports dead load, live load from the moving vehicle, earth pressure on the side walls, hydrostatic pressure from the inside and finally pressure on the bottom slab, in addition to the weight of the slab itself [3–5].

The earth exerts active (minimum), passive (maximum), and resting pressure (average pressure), which will depend on the condition of the soil [6]. When the structure is built before the backfill is placed and the structure is not in a position to yield on either side, the earth pressure will reach a state of rest. In the case of the box, since it is confined with earth on both sides, the earth state must be at rest and a coefficient higher than the active pressure is normally adopted in the design [6].

The earth is filled after construction of the box and the box is not in a position to move/yield, so the pressure should be at rest. It is common practice to take the box as a rigid frame. Also, the unit length of the box is taken to design the effect of all forces acting on the unit length, that is 1.00 m. Moments at center and supports for slabs and walls are obtained for combinations of service load and resistance, and elements are designed for the maximum moment they can be subjected to. While the shear force is evaluated at a distance equal to the effective depth of the wall face [5].

The theory of wheel load distribution through the fill was investigated by several researchers [7, 8], as a result, there are two general solutions available, the closed and numerical form. The characteristics of each approach were briefly outlined in this paper. The closed-form solution treats the soil as a continuum and the solution is given anywhere in the problem domain in the form of equation [9]. This type of solution is complicated to obtain for all types of loading and boundary conditions without making simplifications in stress-strain relationships, directional properties of the soil and geometry of the culvert box, decreasing appreciably the accuracy of the problem.

Finally, the numerical solution incorporates the nonlinearities that the closed-form method does not solve in a finite element model together with a calibration of experimental results, allowing the calibration of finite element models and their extrapolation to models using frame-type elements [10–12]; in the present work they will be used in their modeling. Different works have been carried out in Colombia, ranging from pre-design analysis and design [13], as well as pathological studies in culvert boxes [14], among others.

The importance of taking into account the different physical actions on the box culverts is reflected in the magnitude of the cutting forces and the bending moments generated by these physical actions. In addition, this research established a comparison between the amounts of concrete volumes and kilograms of steel for different backfill heights using the American Association of State Highwayand Transportation Officials standard (AASHTO) [5] and Colombian bridge code (CCP14) [3].

The numerical simulation was performed in SAP2000 [15] and the design was carried out using the Visual Basic programming language [14]. The objective of this research was to design box culverts for different fill heights using the [3] and [5], as a result it was found that the AASHTO [5] requires larger volumes of concrete than the CCP14 [3].

2. Materials and methods

This research work determined the box culvert designs of 2.00 m × 2.00 m and 2.50 m × 2.50 m since these are the most commonly used sections in Colombia with different fill heights (H). Figure 1 shows truck axle, tandem and their respective loads used in the design of box culverts and defined below.

Table 1 shows the geometric dimensions and the analysis and design parameters used in this work; the note below Table 1 indicates the difference between the two box culverts under study. For the tandem axle, the expressions are the same as for the design truck, only the variables $S_a = 1200$ mm and $P = 125$ kN change. A schematic representation of the loads used for the analysis performed in the SAP2000 software [15] and design are shown in Figure 2.

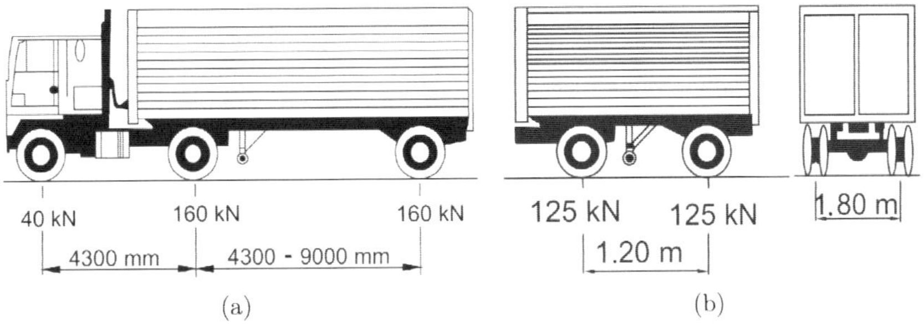

Figure 1. Design vehicle configuration, (a) truck and (b) tandem.

Table 1. Design parameters for box culvert 2.00 m × 2.00 m and fill height of 5.00 m

Variable	Value
Concrete compressive strength, f'_c (MPa)	34.47
Yield strength of steel, f_y (MPa)	413.69
Unit weight of concrete, γ_c (kN/m)	23.20
Unit weight of soil, γ_s (kN/m^3)	19.25
Clear height of box culvert, H_l (mm)	2000
Free base of box culvert, B (mm)	2000
Wall thickness, T_s (mm)	200
Thickness of top slab, T_l (mm)	180
Thickness of bottom slab, T_b (mm)	200
Backfill height, h_f (mm)	5000
Soil friction angle, (degrees)	30

Note: thickness of the upper slab is 230 mm and the thickness of the lower slab is 250 mm.

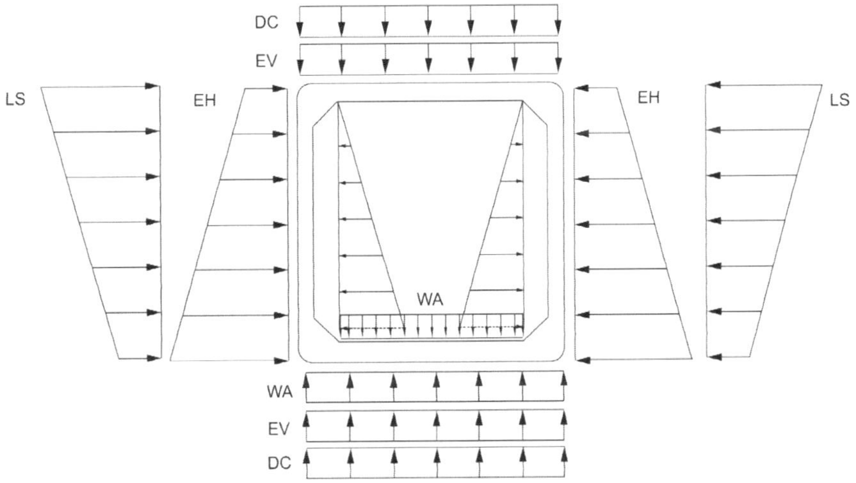

Figure 2. Compressive strength envelope of solid clay brick masonry units.

The design combinations used for both the CCP14 [3] and AASHTO [5], for the strength limit state and the serviceability limit state. For the resistance limit state condition, three combinations are identified: Ia, Ib and Ic. Combination Ia takes into account the factored loads; maximum vertical load and maximum horizontal load, Equation (1).

$$1.25DC + 1.30(1.05)EV + 1.75(LL + IM) + 1.35(1.05)EH_{max} + 1.75LS, \quad (1)$$

where DC represents dead load, EV is vertical thrust of the ground, LL corresponds live load, IM represents dynamic load capacity, EH_{max} is maximum earth pressure on the walls and LS corresponds live load overload. Combination Ib takes into account the factored loads; maximum vertical load and minimum horizontal load, see Equation (2).

$$1.00DC + 1.00EV + 1.00(LL + IM) + 1.00WA + 1.00EH_{min}, \quad (2)$$

where WA represents water load and EH_{min} is minimum horizontal load. Combination Ic takes into account the factored loads; minimum vertical load and maximum horizontal load, see Equation (3).

$$0.90DC + (0.90/1.05)EV + 1.35(1.05)EH_{max} + 1.75LS. \quad (3)$$

The serviceability limit state condition, also has three combinations are identified: Ia, Ib and Ic. Combination Ia takes into account the unfactored loads; maximum vertical load and maximum horizontal load, see Equation (4).

$$1.00DC + 1.00EV + 1.00(LL + IM) + 1.00EH_{max} + 1.00LS. \quad (4)$$

Combination Ib takes into account the unfactored loads; maximum vertical load and minimum horizontal load, see Equation (5).

$$1.00DC + 1.00EV + 1.00(LL + IM) + 1.00WA + 1.00EH_{min}. \quad (5)$$

Finally combination Ic takes into account the unfactored loads; minimum vertical load and maximum horizontal load, see Equation (6).

$$1.00DC + 1.00EV + 1.00EH_{max} + 1.00LS. \quad (6)$$

3. Results and discussions

This was performed from the numerical model made in [15] and taking into account the envelope of results for bending moments and shear force. Figure 3 shows the envelope of moments (kN-m) for the strength limit state, see Figure 3(a) and serviceability limit state, see Figure 3(b) using the CCP14 [3].

The moment diagrams presented in Figure 3(a) have a expected behavior and its maximum positive and negative moment values are as expected, likewise the results obtained in Figure 3(b). The results obtained from the analysis of loads, moments and shear forces were processed for design purposes in the Visual Basic programming language. The results obtained were compared using the CCP14 [3] and AASHTO [5] codes, as well as the document of the "Ministerio de Transporte", Colombia [16]. The design concrete volumes for a 2.00 m × 2.00 m and 2.50 m × 2.50 m box culvert with free discharge length and fill heights of: 0.60 m, 1.00 m, 2.00 m, 3.00 m, 4.00 m and 5.00 m are shown in the Figure 4.

Figure 4(b) presents differences due to the minimum thickness requirements defined in the CCP14 [3] and AASHTO [5], the design load and the clear span between walls; this did not generate changes in the wall and slab thicknesses for the different fill heights. The concrete volumes using [16–18] do not present variations in the thicknesses for fill heights ≤ 3.00 m, although they are less than those found in the two previous codes; however, for fill heights 3.00 < H ≤ 5.00 m, the concrete volumes increase, becoming equal to those found using CCP14 [3]. The difference in concrete volume/m (per meter of box culvert) for the two codes is due to the fact that AASHTO [5] requires the use of the four haunches (diagonal elements located at the four corners of the box), while CCP14 [3] does not.

A similar analysis was performed for box culverts of 2.50 m × 2.50 m and variable fill height; the results can be seen in Figure 4(a). The difference in quantities of m^3 of concrete per meter of box culverts shown in Figure 4(a) is due to the presence of the haunch in the design using [5], although this difference in volumes is slightly reduced due to the increase in wall thickness in the design using [3]. The technical regulation presents lower quantities of concrete up to a fill height of 4.15 m with respect to CCP14 [3], and does not meet the minimum requirements established in [3].

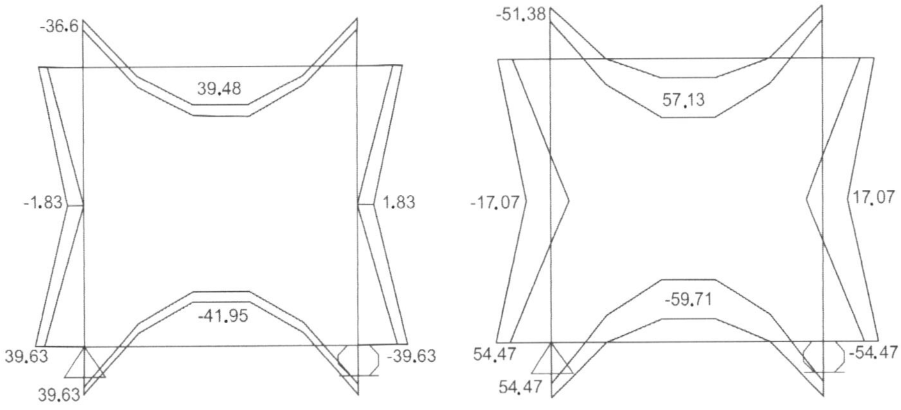

Figure 3. Moment envelope for a section of 2.00 m × 2.00 m and a fill height of 5.00 m, for (a) combination by serviceability and (b) combination by strength.

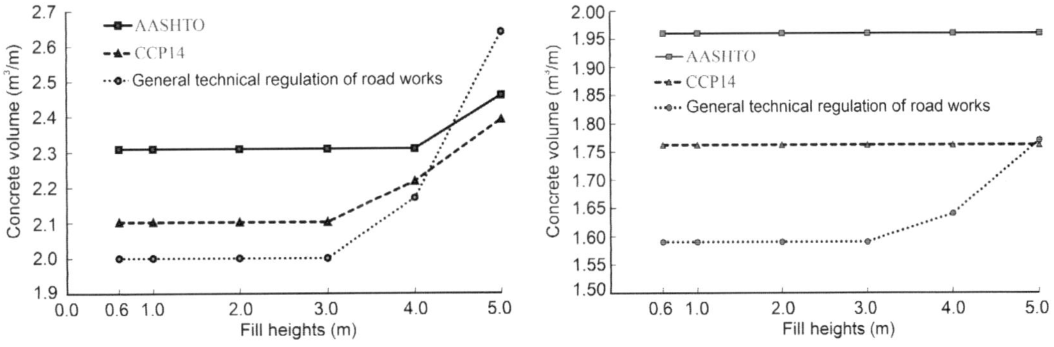

Figure 4. Comparison of concrete volumes for culverts box of (a) 2.50 m × 2.50 m and (b) 2.00 m × 2.00 m.

An analysis of the amount of kg/m of culvert box for a section of 2.00 n x 2.00 m free and variable fill height is shown in Figure 5. The steel requirements in kg/m box culverts present in Figure 5(b) are higher in CCP14 [3] than all other papers under study, for fill heights ≤ 2.00 m. For heights > 2.00 m both codes are similar, and for all fill heights [9] has less reinforcing steel performance. The differences between the [3] and [5] are because in AASHTO [5] the moments are taken at the beginning of the haunch, while in CCP14 [3] the bending moments are evaluated in the axis of the box wall thickness, where the moments increase even more due to the presence of the dynamic load. This generates an important difference in the design moments due to the increased stiffness provided by the haunch.

The behavior curve of the reinforcing steel in kg/m for the 2.50 m × 2.50 m box culverts shown in Figure 5(a) is very similar to the one found for the 2.00 m × 2.00 m box culverts, influenced by the presence of the haunch in the AASHTO [5].

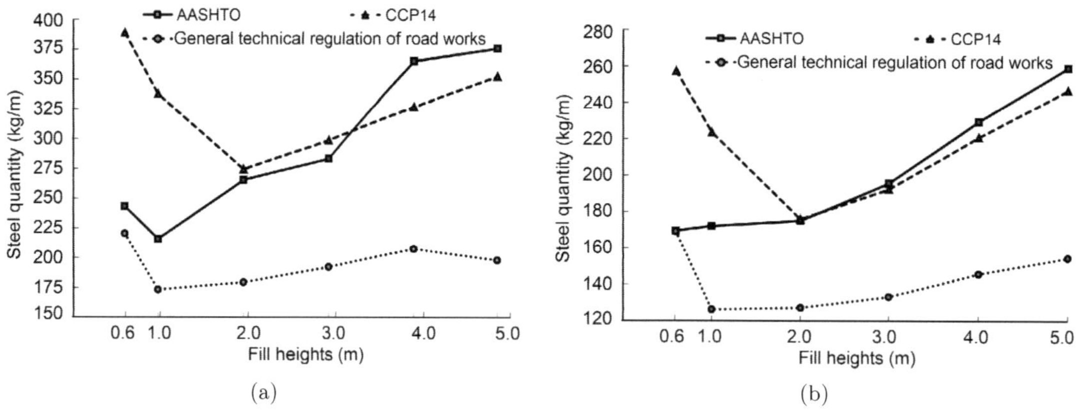

Figure 5. Comparison kilograms of reinforcing steel for culverts box of (a) 2.50 m × 2.50 m and (b) 2.00 m × 2.00 m.

4. Conclusions

This research work makes a contribution to the design of box culverts taking into account different physical actions such as earth fill load, live load, over live load, dead load, water pressure and impact load, as well as different load combinations that seek to increase the effects on the structure. The Colombian bridge standard estimates lower quantities of cubic meters of concrete than its similar one from the American Association of State Highway and Transportation Officials standard, and the use of the old general technical regulation of road works provides for lower quantities of concrete, possibly due to ignorance of the behavior and the demand for resistance for the road works. years 1988. For steel, the Colombian bridge standard has greater requirements for fills of less than 3.00 meters, while for fills greater than 3.00 m, the American Association of State Highway and Transportation Officials standard has greater steel requirements.

References

[1] Xanthakos P P 1994 *Theory and Design of Bridges* (Ottawa: John Wiley & Sons)
[2] Mackie K, Stojadinović B 2003 *Seismic Demands for Performance-Based Design of Bridges* (Berkeley: Pacific Earthquake Engineering Research Center)
[3] Asociación Colombiana de Ingeniería Sismica (AIS) 2014 *Norma Colombiana de Diseño de Puentes LRFD-CCP14* (Bogotá: Asociación Colombiana de Ingeniería Sismica)
[4] Vallecilla C R 2006 *Manual de Puentes en Concreto Reforzado* (Bogotá: Universidad Nacional de Colombia)
[5] American Association of State Highway and Transportation Officials (AASHTO) 2012 *LRFD Bridge Design Specifications* (Washington DC: American Association of State Highway and Transportation Officials)
[6] Terzaghi K 1962 *Theoretical soil Mechanics* (London: John Wiley and Sons)
[7] Spangler M G, Handy R L 1982 *Soil Engineering* (New York: Harper and Row)
[8] Burmister D M 1956 Stress and displacement characteristics of a two-layer rigid base soil system: influence diagrams and practical applications. *Proceedings of the Thirty-Fifth Annual Meeting of the Highway Research Board* (Washington DC: Highway Research Board) p 773
[9] Poulos H G, Davis E H 1974 *Elastic Solutions for Soil and Rock Mechanics* (New York: John Wiley and Sons)

[10] Abdel-Karim A M, Tadros M K, Benak J V 1993 Structural response of full-scale concrete box culvert *Journal of structural Engineering* **119(11)** 3238
[11] Sinha B N, Sharma R P 2009 RCC box culvert-methodology and designs including computer method *Journal of the Indian Roads Congress* **10(555)** 189
[12] Kolate N, Mathew M, Mali S 2014 Analysis and design of RCC box culvert *Int. J. Sci. Eng. Res.* **5** 36
[13] Gualpa Chacón L A 2018 *Evaluación y Rediseño de una Alcantarilla de Hormigón Armado Ducto Cajón, en el Barrio Bellavista del Cantón La Troncal* (Guayaquil: Universidad Laica Vicente Rocafuerte)
[14] Muñoz Barrios D P 2017 *Estudio Patológico a Cuatro Box Culverts Av. Boyacá Parafutura Ampliacion y/o Adecuación para Sistema de Transporte Masivo Transmilenio* (Bogotá: Universidad Santo Tomás)
[15] SAP2000 V22 2020 *Structural Analysis Program* (Washington DC: Computer and Structures Inc.)
[16] Ministerio del Transporte 1988 *Reglamento Técnico General de Obras Civiles, Tomo VI* (Colombia: Ministerio del Transporte)
[17] Afanador G N, Prada Y S, Cáceres D M 2013 Diseño sismico de muros de contención en gravedad y voladizo *Revista Colombiana de Tecnologias de Avanzada* **2(20)** 97
[18] Afanador G N, Carrascal Delgado M, Bayona Chinchilla M J 2013 Experimentación, comportamiento y modelación de la tapia pisada-caso de estudio: complejo histórico Gran Convención *Revista Facultad de Ingeniería* **22(35)** 47

Structural and physical evaluation of a reinforced beam using strain gauges

N Afanador-Garcia[1], G Guerrero-Gomez[2], and R Gallardo-Amaya[3]
[1] Civil Engineering Research Group, Universidad Francisco de Paula Santander, Seccional Ocaña, Colombia
[2] Grupo de Investigación en Tecnología y Desarrollo en Ingeniería, Universidad Francisco de Paula Santander, Seccional Ocaña, Colombia
[3] Grupo de Investigación en Geotecnia y Medio Ambiente, Universidad Francisco de Paula Santander, Seccional Ocaña, Colombia

E-mail: nafanadorg@ufpso.edu.co, gguerrerog@ufpso.edu.co

Abstract. In this research project, the deformations in the longitudinal and transverse reinforcing steel of a reinforced concrete beam with 2ϕ½" were estimated. Additionally, the displacements in the center of the span were measured together with the loads, which generated the cracking of the beam of dimensions 180 mm×240 mm×3100 mm. Displacement were performed using a linear variable displacement transducer, and strain gauges were used to measure deformations. Finally, the applied load's measurements were obtained with a load cell Pinzuar/Model-100/20T equipment. The physical-mechanical properties of the concrete used were determined through compressive strength tests at 28 days and modulus of elasticity. For this purpose, a load-bearing frame was used to support the beam at three points for load application. The data was collected directly on the Quantum/X2 equipment and analyzed with the help of the Catman/AP software. The maximum deformations found in the bending test at three points in the reinforced concrete beam were not greater than 8483 μm/mm, presenting a failure in the center of the beam due to the creep of the tensile steel for a maximum load of 3115 Kg. Finally, the physical behavior of the reinforced concrete beam with applying a load allows evaluating and optimizing this kind of systems.

1. Introduction

Strain gauges are piezoresistive sensors, manufactured with metallic or semiconductor materials (silicon or germanium). The main objective of this type of sensors is to determine the deformations in any direction or sense in a specific point of a structure that can be metallic, non-metallic or concrete, considering the effect of the electrical resistance that the strain gauge has with a micro-measurement, which works under the effect of Hooke's law, which states that the deformation of an elastic material is directly proportional to the applied normal force [1]. Thus, strain gauges are widely used in civil engineering because it is possible to predict the deformations, stresses, and cracks generated in structural elements of any material and type of element (beams, walls, masonry units, among others) [2].

Considering the literature review, several studies have addressed the analysis of beams of different materials in order to determine the loads where the material generates creep and thus the initiation and propagation of cracks that induce the failure of the material and/or structure. According to Masri, *et al.* [3], fiber optic sensors or strain gauges integrated to structures provide valuable information about the state of deformations of different materials when subjected to loads through monitoring during the test. Moreno-Robles and García [4] determined the loads applied to a train rail using strain gauges. To validate the results and the data obtained correctly, the authors performed a 3D numerical model using FLAC3D software [5]. The results showed that the strain gauges must be separated at least 50 mm from the point where the load is applied to obtain results that match the real values. Huang, *et al.* [6] evaluated the properties of concrete based on the optical technique automatic digital image. correlation (ADIC) using strain gauges. This correlation process helps to automatically determine the deformation of concrete in real time without the need for visual inspection.

The results demonstrated that it is possible to propose a new optical strain gauge to monitor in real time the strain and determine the compressive modulus of elasticity of concrete based on the ADIC model, being an accurate, cost-effective, and reliable method. Kim, *et al.* [7] developed a new silicon strain gauge using microelectromechanical systems (MEMS) technology. The results showed that the proposed gauge minimizes the position change in measurement and improves the bond strength thus improving the sensor performance and in turn the manufacturing cost. Payo and Feliu [8] designed a sensor system to determine real-time 3D deformations of flexible beams using strain gauges. The results showed that the designed system meets the specifications to obtain quality measurements that contribute to the analysis of the results with few computational requirements. Finally, Chen, *et al.* [9] developed a strain gauge design to obtain the strain distribution in a region of a structure. The results show that the proposed design is reliable and robust for different industrial applications.

Considering the above, the present research work aims to estimate the deformations in the longitudinal and transverse reinforcing steel of a reinforced concrete beam with 2 ½" up and down of a beam of dimensions 180 mm×240 mm×3100 mm using 10/120-CLY41-3L-0.5M strain gauges. The assembly was carried out on a load frame with a capacity of 50 t in compression and 30 t in tension, and the data were extracted using Quantum/X2 equipment with the help of Catman/AP software [10].

2. Methodology and materials

Strain gauges are widely used to determine length changes in different parts of a structure. Figure 1 shown the typical geometry of a strain gauges used to the developed of different studies in reinforced concrete (beams) [11]. The commercial stain gauges had the following parameters, type 10/120-CLY41-3L-0.5M, resistance 120 Ω, K gauge of 2.11, and transverse sensitivity 0.2%. Notice that the resistance of the strain change with the gauge deformation [11].

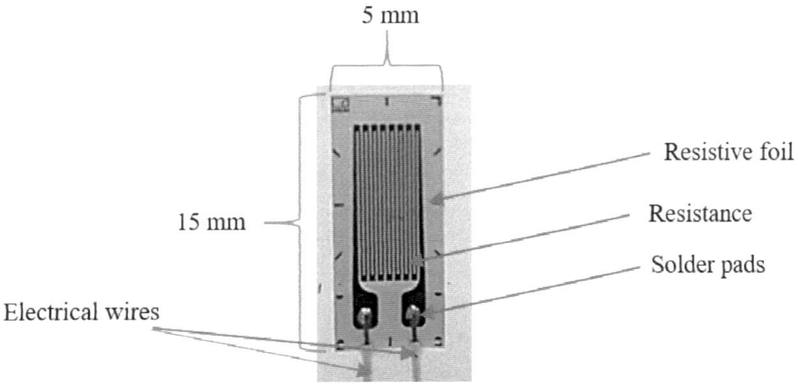

Figure 1. Main characteristics of a strain gauge.

The installation of the gauge on the rods, it is very important to have a smooth and clean surface that allows a proper installation and adhesion to ensure a correct measurement of the values that can be obtained during the test [12]; the gauge must be well installed to ensure adhesion with the surface, and for this, the procedure described in Figure 2 was considered.

Journal of Physics: Conference Series **2153** (2022) 012003 doi:10.1088/1742-6596/2153/1/012003

Considering the above, the 180 mm×240 mm×3100 mm reinforced concrete beam was fabricated considering the configuration shown in Figure 3; the rods used were 2 ½ inch rods top and bottom with 9 $^{3}/_{8}$ inch stirrups every 60 mm and in the center every 120 mm. The strain gauges were installed considering the steps in Figure 2, and located in the beam as detailed in the red circles, one pair at the end, and another pair in the center of the beam where the maximum stress and deformation is generated due to the load applied during the development of the test [13].

From step 4, presented in Figure 2 (set-up and execution of the test), the process for the execution of the laboratory test; initially, a load frame was used where the instrumented beam was supported to perform the three-point bending test. Subsequently, after the application of the gradual load at the midpoint of the beam, using a Pinzuar/Model-100/20T load cell, deflections were generated, which were obtained with the help of the REP/LDT 50 mm and 100 mm linear variable displacement transducer (LVDT) equipment with 2 mV/V output; the data were continuously recorded in the Quantum/X2 equipment and later analyzed in the Catman/AP software [10].

Setting the direction of the strain gauges on the rod required a study to determine the point to be evaluated, the main coordinate planes and the type of mechanical stress (tensile or compressive) to which the structure (beam) will be subjected; there are two types of applications that gages can have. The first one is because of the variable to be found, for example the deformation, which is intermediate; the second is the state to be known (tension or compression). Strain gauges are useful for all those situations in which it is necessary to find stresses and strains in structures that obey Hooke's law.

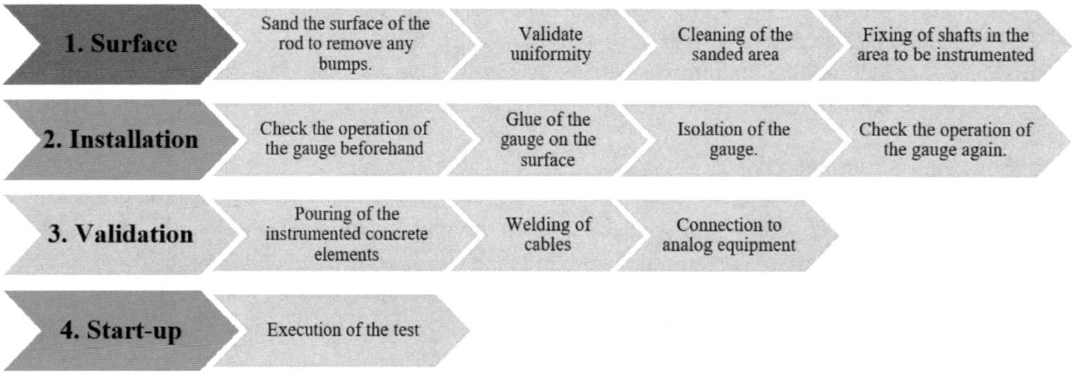

Figure 2. Guidelines applied to develop tests.

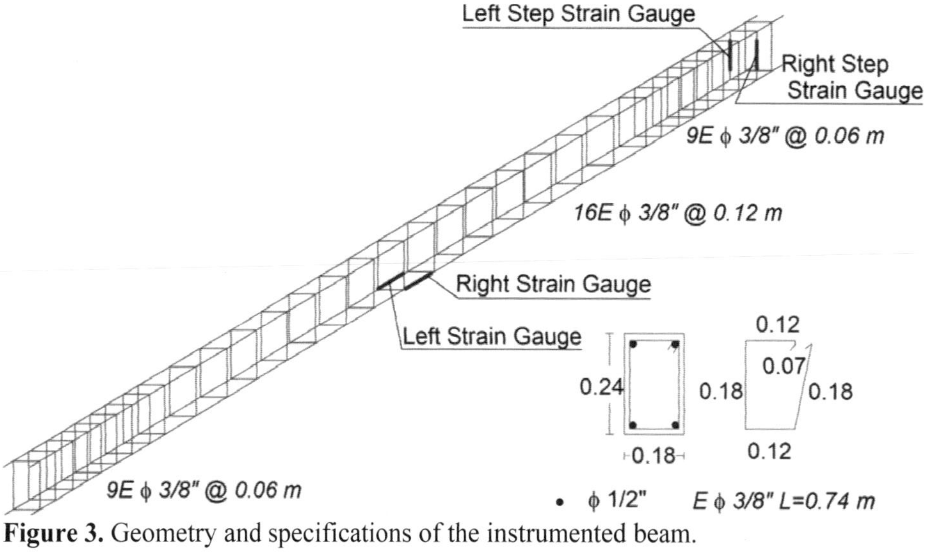

Figure 3. Geometry and specifications of the instrumented beam.

3. Results and discussions

In this section, the main results collected from the flexural test on a beam with strain gauges are reported, considering from the properties of the concrete until the analysis of the results. The parameters and physical-mechanical properties using of the concrete used for the manufacture of beams reinforced with rebar are as follows: dosage 1:2:3, compressive strength at 28 days = 9.05 MPa, slump 11.5 mm, density 2354.67 Kg/m3, and modulus of elasticity 1.00 MPa.

The reinforced concrete beam with the strain gauges incorporated in the rods was instrumented with its respective mounting in the load frame, as shown in Figure 4; the beam with its respective connections and the load cell, whose information during the development of the test was collected as detailed in Figure 4. The configuration used for the evaluation of the reinforced beam in the three-point bending test, using strain gauges, is similar to that reported by [14-16].

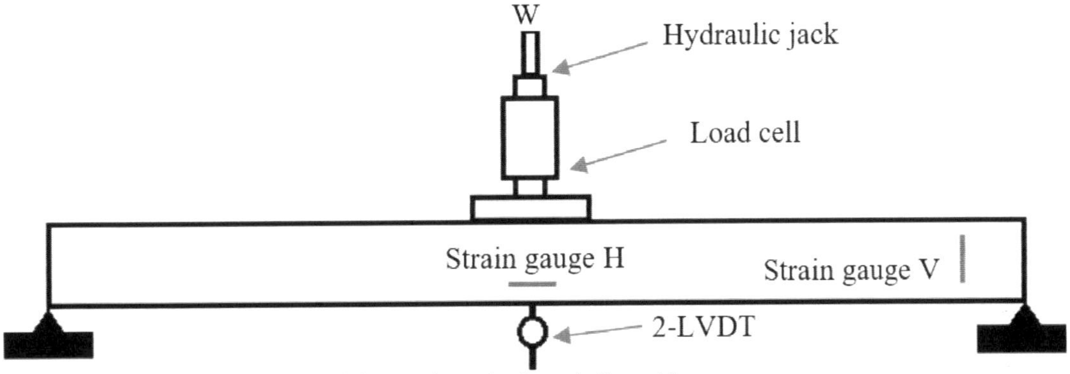

Figure 4. Schematic diagram of flexural test in the reinforced beam.

Once the flexural test on the beam was completed, the results were compiled and analyzed in the Catman/AP software [10], obtaining the results shown in Figure 5. Figure 5(a) shows the behavior of the load applied using a hydraulic jack and measured through the load cell, with a load speed of approximately 170 Kg/min, which increased up to 3200 Kg. This is the instant at which the reinforcing steel creep and the flexural failure of the reinforced concrete beam occurred. Note that, t is the test time, P is the applied load, Δ is the displacements and Ɛ are the deformation.

Figure 5(b) indicates the behavior of the average deflection, that is, the average value for the two LVDTs located in the center of the beam, in order to avoid distortion in the beam deflection; the observed behavior indicates a small load adjustment zone in the beam due to a small preload. The deflection-time relationship is linear, with a deflection rate of 0.020 mm/s, after a load equal to 180 Kg is applied.

Figure 5(c) indicates the deflection behavior of the average longitudinal reinforcement, that is, the average of the two gauges installed on the 2 ½” steel bars located at the bottom of the beam, in order to avoid distortion in the deformation; the behavior is linear up to the theoretical creep deformation of 0.0021, while the experimental creep deformation was 0.0027. Figure 5(d) shows how the average deformation varies over time in a stirrup located 60 mm from the beam support, the strain gauges located on the vertical sides of the stirrup are intended to avoid distortion in the deformation.

Due to the effect of the load applied on the beam at the center of the span, this element failed by creep of the bottom steel, generating the presence of cracks from the bottom to the top of the beam as shown in Figure 6. It was observed that the loads that generated cracking were between 255 Kg and 3115 Kg, the latter when the beam failed.

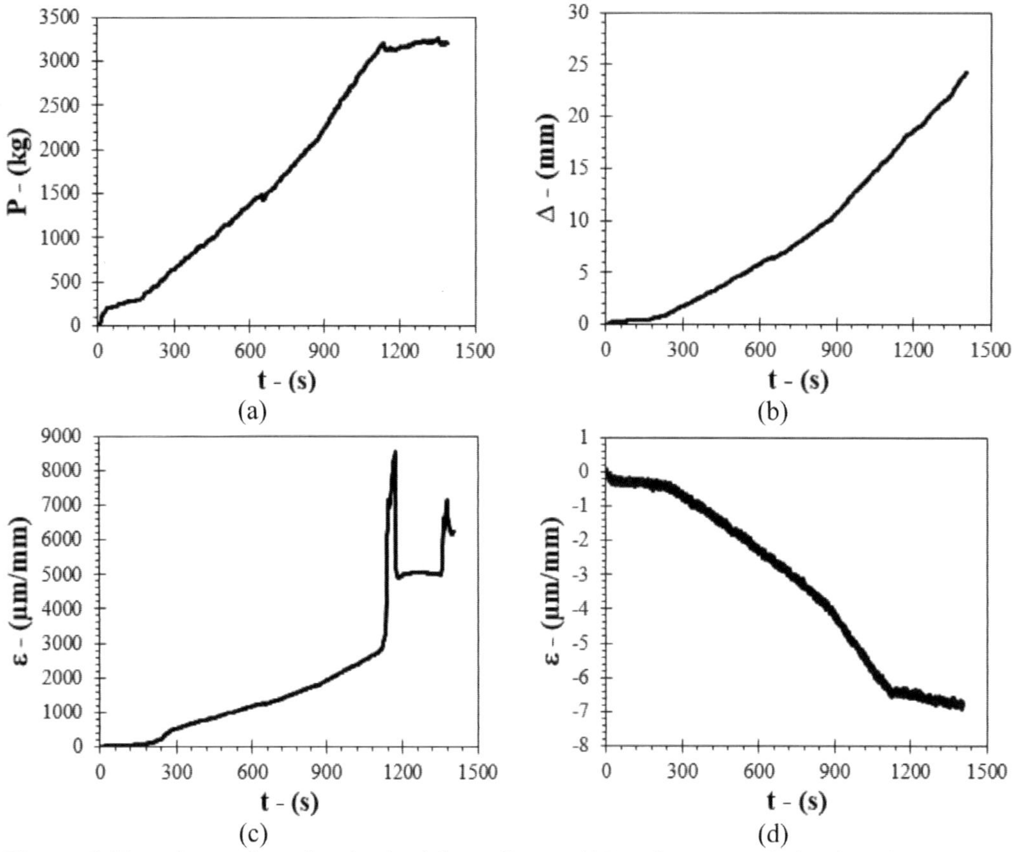

Figure 5. Experimental results obtained from Catman/AP software; (a) behavior of the applied load, (b) average vertical displacement at the center of the beam, (c) average deformation of longitudinal reinforcement, (d) average deformation of transverse reinforcement.

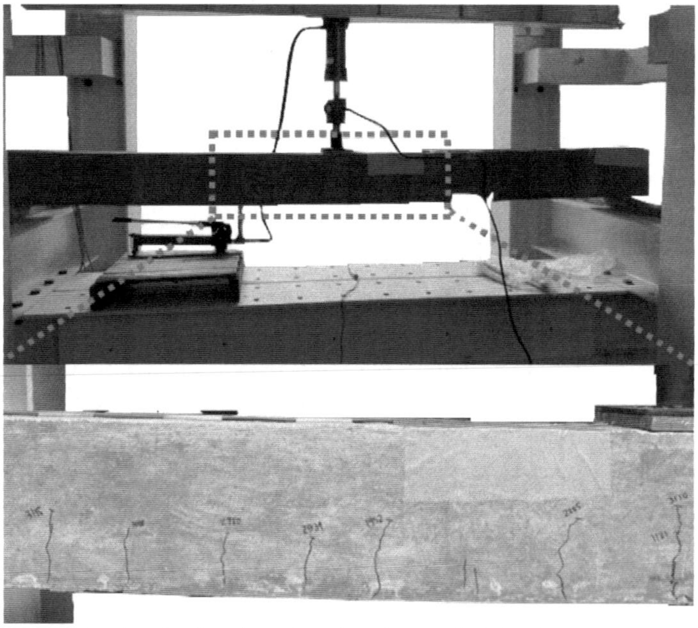

Figure 6. Cracks developed under the applied load.

4. Conclusions

It was established that the theoretical statement that creep occurs in a smoothed form is not true; there is a rearrangement of the grains that make up the reinforcing bars at the moment of creep and there is a small jump in creep deformation for small variations in load. After steel creep, there is a relaxation zone where the deformation increases without load increase.

The maximum deformations found in the bending test at three points in the reinforced concrete beam were not greater than 8350 μm/mm, presenting a failure in the center of the beam due to the creep of the tensile steel for a maximum load of 3115 Kg. From the analysis developed for the cracks generated under the applied load it can be seen that there are no significant shear cracks, due to the small separation between stirrups that does not allow the shear cracking to propagate, contrary to that presented by bending that they distribute along and above of the beam.

References

[1] Hannah R L, Reed S E 1992 *Strain Gage Users' Handbook* (London: Chapman and Hall)
[2] Hoffmann K 1989 *An Introduction to Measurements using Strain Gages* (Darmstadt: Hottinger Baldwin Messtechnik GmbH)
[3] Masri S F, Agbabian M S, Abdel-Ghaffar A M, Higazy M, Claus R O, de Vries M J 1994 Experimental study of embedded fiber-optic strain gauges in concrete structures *Journal of Engineering Mechanics* **120(8)** 1696
[4] Moreno-Robles J, García-De La Oliva J 2012 Determinación de las cargas aplicadas al carril mediante bandas extensométricas *Ingeniería Civil* **165** 123
[5] Itasca Consulting Group 2020 *Fast Lagrangian Analysis of Continua in 3 Dimensions, FLAC3D V7.0* (Minneapolis: Itasca Consulting Group, Inc.)
[6] Huang Y H, Liu L, Sham F C, Chan Y S, Ng S P 2010 Optical strain gauge vs. traditional strain gauges for concrete elasticity modulus determination *Optik* **121(18)** 1635
[7] Kim J W, Kim K B, Kim J H, Min N K 2018 Bulk-micromachined, SOI-based half-bridge silicon strain gauges for high pressure applications *Journal of Micromechanics Microengineering* **28(1)** 125004:1
[8] Payo I, Feliu V 2014 Strain gauges based sensor system for measuring 3-D deflections of flexible beams *Sensors and Actuators A: Physical* **217** 81
[9] Chen X, Topac T, Smith W, Ladpli P, Liu C, Chang F K 2018 Characterization of distributed microfabricated strain gauges on stretchable sensor networks for structural applications *Sensors* **18(10)** 3260
[10] Hottinger Baldwin Messtechnik (HBM) 2020 *Catman V 5.5.1* (Darmstadt: Hottinger Baldwin Messtechnik)
[11] Regier R, Hoult N A 2014 Distributed strain behavior of a reinforced concrete bridge: case study *Journal of Bridge Engineering* **19(12)** 05014007:1
[12] Schaefer P L, Barrier G, Chagnon G, Alonso T, Moreau-Gaudry A 2019 Strain gauges based 3D shape monitoring of beam structures using finite width gauge model *Experimental Techniques* **43(5)** 599
[13] Forero J 2006 Strain gages. Fundamentos y aplicaciones en resistencia de materiales *Avances: Investigación en Ingenierías* **4** 37
[14] Afanador García N 2019 Quantification of uncertainty in metallic elements subjected to fatigue *Journal of Physics: Conference Series* **1329(1)** 12011:1
[15] Gunasekaran K, Annadurai R, Kumar P S 2013 Study on reinforced lightweight coconut shell concrete beam behavior under flexure *Materials and Design* **46** 157
[16] Afanador García N, Guerrero Gómez G, Nolasco Serna C 2021 Reinforced concrete beams subjected to three-point bending using finite element method *Journal of Physics: Conference Series* **1938(1)** 012008:1

An application of physics: simply supported bridges made of post-tensioned concrete and structural steel beams

J F Márquez-Peñaranda[1], J R Pineda-Rodríguez[1], and J P Rojas-Suárez[1]
[1] Universidad Francisco de Paula Santander, San José de Cúcuta, Colombia

E-mail: jorgefernandomp@ufps.edu.co

Abstract. Bridges represent an important application of physics capable of solving real transportation problems. Knowledge of convenience of different mechanical solutions when analyzing and designing bridge is needed. For these reasons, this work is focused on the study of convenience of using two types of bridges. Simply supported short-medium span bridges (30 m to 45 m) are usually excessively long when choosing reinforced concrete solutions and usually short for other types of structures such as cable-stayed or cantilever bridges. The suitability of simply supported bridges leads to the need of studying their cost benefit ratios. This work studies the cost benefit ratio for post-tensioned concrete beams and structural steel girders in simply supported straight bridges. Eight models built of type I sections were used in both cases to analyze the bridges using a software based on the stiffness method. Span of each bridge was set to 30 m, 35 m, 40 m, and 45 m. The convenience of each type of bridge was done comparing the total and the cost per linear meter of each solution (post-tensioned and structural steel). Comparison was done using material consumption, labor, and construction processes costs only. Also, allowable vertical displacement given by current bridge design standards was verified.

1. Introduction

Bridges are an important application of physics for solving problems of road infrastructure works. Structural configuration of bridges depends on the natural or artificial obstacles to overcome and their characteristics such as lengths, shape, and height. There are numerous types of structural systems for bridges made of different materials such as reinforced concrete, post-tensioned concrete and structural steel are used [1]. To guarantee comparable safety and functionality of bridges methodologies for analysis and design of bridges are stated in worldwide standards. Some of the most important standards are American Association of State Highway and Transportation Officials (AASHTO), Load Resistant Factor Design (LRFD), bridge design specifications [2]; American Concrete Institute (ACI), ACI 318S-14 [3]; "Asociación Colombiana de Ingeniería Sísmica (AIS)", LRFD-CCP 14 [4] and NSR-10 [5].

Structural design of a bridge must consider the economic feasibility without prejudice to its functional purposes [6]. Post-tensioned concrete and structural steel are the most common materials for simply supported short-medium span bridges having lengths between 30 m and 45 m. When dealing with the construction of bridges, post-tensioned concrete or structural steel are widely used as an economic structural system, although this depends on the bridge length and the amount of supports [7,8].

Several studies have been carried out in relation to the study of costs of design and construction of bridges. Haas [9] investigated the cost effectiveness of steel girders compared to conventional reinforced concrete girders used in bridge construction in South Africa. Similarly, Delgado and Zuñiga [10] reported a cost comparison between bridges with steel girders, reinforced concrete and post-tensioned considering the variation of the free length of the span. On the other hand, Almeida and Armas [11] studied the economic behavior of two types of bridges, comparing the superstructure of 30 m span bridges supported only in two ends and considering post-tensioned concrete beams and gantry beams, designed according to some international bridge design standards. However, there has not been found information reported for Latin-American countries related to the scope of the present study.

The purpose of the present investigation is to compare the cost of two types of materials when used in the construction of simply supported short-medium span bridges using lengths of 30 m, 35 m, 40 m, and 45 m. To do so, bridges designed with structural steel beams and post-tensioned concrete beams are studied to determine which of the solutions has the best economic viability while maintaining a similar level of functionality.

2. Methodology

The research was carried out around the economic and functional study of medium-short vehicular bridges, with lengths between supports of 30 m, 35 m, 40 m, and 45 m. The bridge deck, for all cases, was designed using a reinforced concrete slab with a lane width of 8 m. To involve the most used materials used in practice, in this study post-tensioned concrete and structural steel were used. Safety and functionality were considered to define comparative costs. Safety computation was based on to beams strength while functionality was related to allowable vertical displacements.

2.1. Determination of materials quantities from optimized structural design

Post-tensioned concrete beams were conformed using I-shaped sections. As shown in Figure 1, four post-tensioned beams were outlined to support a typical 18 cm-thick reinforced concrete deck. The stiffness method was used to obtain results for comparison. Figure 1 shows an example of the typical configuration of the bridge section. Dimensions of each post-tensioned concrete beam vary according to the free span of the bridge. The larger the span the bigger the dimensions of the beam section. Medium flange width (B) and total height (H) of beam section of were specified according to each case. Values of B = 0.75 m, 0.80 m, 0.93 m, and 0.95 m were set for span beams of 30 m, 35 m, 40 m, and 45 m, respectively. Similarly, values of H = 1.80 m, 1.95 m, 2.15 m, and 2.35 m were set for span beams of 30 m, 35 m, 40 m, and 45 m respectively.

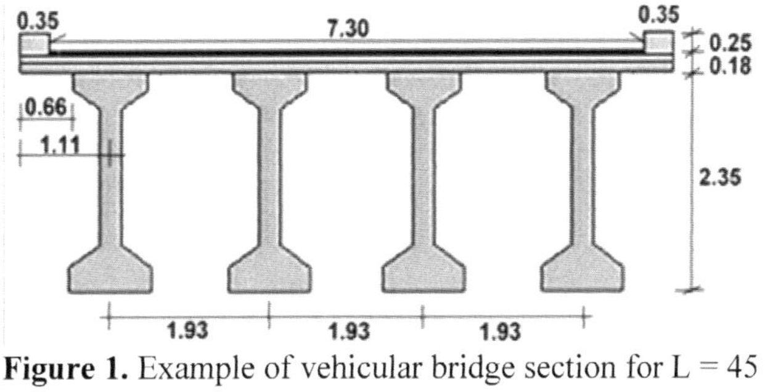

Figure 1. Example of vehicular bridge section for L = 45 m with reinforced concrete deck supported on post-tensioned concrete beams (*Dimensions in meters).

Each structural steel beam was designed using I-shaped section. Four steel beams were outlined to support a typical 18 cm-thick reinforced concrete deck. The stiffness method was used to obtain results for comparison. Figure 2 shows an example of the typical configuration of the bridge section. Dimensions of each steel beam vary according to the free span of the bridge.

The larger the span the bigger the dimensions of the beam section. Flange width (B) and total height (H) of beam section of were specified according to each case. Values of B = 0.40 m, 0.45 m, 0.48 m, and 0.53 m were set for span beams of 30 m, 35 m, 40 m, and 45 m respectively. Similarly, values of

H = 1.50 m, 1.65 m, 1.80 m, and 2.00 m were set for span beams of 30 m, 35 m, 40 m, and 45 m, respectively. Optimized structural design of each solution (post-tensioned concrete or structural steel beam) was carried out computing the minimum materials amount required for guaranteeing safety and functionality conditions stated in the Colombian standards [4].

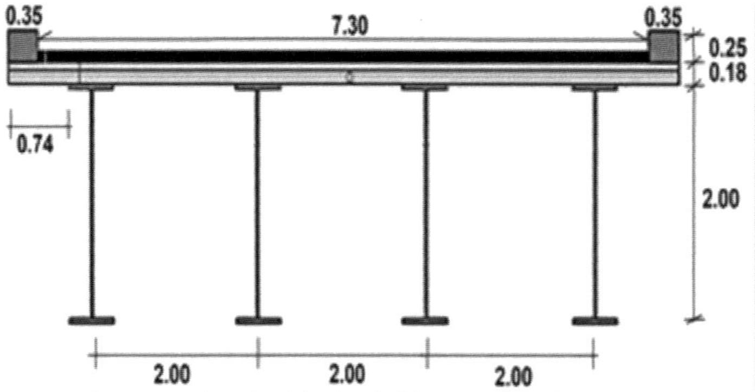

Figure 2. Example of vehicular bridge section for L = 45 m with reinforced concrete deck supported on structural steel beams. (*Dimensions in meters).

2.2. Determination of basic costs of materials
Unit price analysis was carried out for materials required from the results of the structural design of each solution. The costs were based on commercial prices, and they were computed using Colombian pesos (COP) of the year 2021. Costs of material, labor, and assembly of processed materials arisen from the construction of the studied bridges were considered when computing each unit price. Measurement units were based on the international system of units, being cubic meter (m^3), linear meter (m), and kilogram (Kg) the typical units.

2.3. Estimation of the trend line equation
To present results in a practical form, a descriptive equation which relates the trend of costs for each solution was estimated. Least squares method was used to get the best fit for costs associated to simply supported vehicular bridges designed using post-tensioned and structural steel beams. The least squares method calculates from the number of even numbers of experimental data (x,y), the slope m and intersect b values that best fit the data in one straight line. The expression is based on the equation of the line $y = mx \pm b$ (where m is the slope of the line and b is the cut-off point). Equation (1) and Equation (2) allow to compute the slope and the intersect point with vertical axis of the trend line, respectively [12].

$$m = \left(\frac{n*\sum(x*y)-\sum x*\sum y}{n*\sum x^2-|\sum x|^2}\right), \tag{1}$$

$$b = \left(\frac{\sum y*\sum x^2-\sum x*\sum(x*y)}{n*\sum x^2-|\sum x|^2}\right). \tag{2}$$

After calculating the factors m and b, the values are replaced in the expression $y = mx \pm b$. The expression shown in Equation (3) leads to compute the best adjust of a straight line which describes the trend of cost of vehicular bridges by increasing the length between supports from 30 m to 45 m [12]. This operation has been repeated for each type of bridges designed of post-tensioned concrete and structural steel.

$$y = \left(\frac{n*\sum(x*y)-\sum x*\sum y}{n*\sum x^2-|\sum x|^2}\right)x \pm \left(\frac{\sum y*\sum x^2-\sum x*\sum(x*y)}{n*\sum x^2-|\sum x|^2}\right). \tag{3}$$

3. Results and discussion

Table 1 summarizes the reference amounts used for comparison of physical efficiency of solutions made of post-tensioned concrete; bending moment demand has been computed considering the higher expected forces (loads) over the bridge following the current standards [2-4]. The ratio "demand/weight" has been calculated dividing the bending moment demand by the total weight of the bridge. Rows 3 to 8 of Table 1 refer to the material consumption in bridges made using post-tensioned concrete beams; demand/weight ratios vary from 225 KN×m/KN to 415 KN×m/KN following a parabolic trend in function of the span length which is consistent with the trend of dominant bending moment curve.

Physical effect of load lever and forces configuration seem to impose such second-degree trend; as for material costs, it is evident that concrete 35 MPa, prestressing steel and reinforcing steel are the most important cost makers. In contrast, concrete 28 MPa and safety railings are the lower cost contributors. A rising trend of materials consumption is observed; the larger the span the lower the cost variation between two successive spans.

Table 1. Reference amounts for bridges made using post-tensioned concrete beams.

Reference amount	Bridges 30 m	Bridges 35 m	Bridges 40 m	Bridges 45 m
Bending moment demand (KN×m)	44275	63571	89805	117549
Demand/weight (KN×m/KN)	225	280	341	415
Concrete 35 MPa for beams (m^3)	159	229	323	401
Concrete 28 MPa for braces (m^3)	6	7	8	9
Concrete 28 MPa for decks (m^3)	49	57	65	73
Reinforcing steel 420MPa (KN)	99	121	148	174
Prestressing steel for beams (Kg)	44	64	88	117
Safety railing (m)	60	70	80	90

Table 2 shows the reference amounts for bridges made with structural steel beams; in this case, the Demand/Weight ratios vary from 331 KN×m/KN to 612 KN×m/KN following a parabolic trend in function of the span length which is consistent with the trend of dominant bending moment curve. Again, as explained in previous paragraph. physical effect of load lever and forces configuration seem to impose such second-degree trend. Rows 3 to 8 of Table 1 refer to the material consumption in bridges made using structural steel beams. In this case, A572-01 G-50 steel, C 12×20.7 steel profiles and reinforcing steel are the most important cost makers are. A rising trend of materials consumption is observed; the larger the span the lower the cost variation between two successive spans.

When comparing results for demand/weight ratios from Table 1 and Table 2 it is concluded that, from the point of the physics (mechanics), the steel beams provide greater efficiency than post-tensioned beams. It is true because steel beams, which have a lower weight, can support a similar or larger demand than post-tensioned beams. For example, for 30 m span, one KN of structural weight supports a bending moment demand of 331 KN×m in the case of steel beams while post-tensioned beams only can support 225 KN×m. According to the previous reasonings, mechanical efficiency of steel beams is around 1.5 times that of the post-tensioned beams for all the studied spans.

Table 2. Reference amounts for bridges made using structural steel beams.

Reference amount	Bridges 30 m	Bridges 35 m	Bridges 40 m	Bridges 45 m
Bending moment demand (KN×m)	28898	37354	46895	57914
Demand/weight (KN×m/KN)	331	416	511	612
Concrete 28 MPa for beams (m^3)	49	56.53	64.60	72.68
Steel A572-01 G-50 (Kg)	655	851	1,053	1,317
Profiles C 12×20.7 for braces (Kg)	9	9	9	9
Steel profiles for shear connectors (Kg)	10	12	13	15
Reinforcing steel 420 MPa (Kg)	39	46	52	59
Safety railing (m)	60	70	80	90

Table 3 shows the basic materials costs computed for each material used in the design of the studied bridges given in Colombian pesos (COP) of the year 2021. First column of the table describes the material, second column refers the unit of measurement, and third column presents the cost per unit of reference for each material. The units of measurement are expressed using the International System of Units using cubic meters (m^3) for volume, linear meter (m) for length and (KN) for force. The higher unit cost corresponds to prestressing steel which is about three to six times more expensive than other types of steel. According to this, there must could be an important influence of the type of steel in the cost of each solution.

Table 3. Cost of construction materials for the year 2021.

Building material	Unit of measurement	Cost (thousands of COP)
Concrete 28 MPa	m^3	1056
Concrete 35 MPa	m^3	1132
Reinforcing steel 420 MPa	KN	1060
Prestressing steel	KN	6789
Steel A572-01 G-50	KN	2080
Steel profiles C 12×20.7	KN	1794
Steel profiles	KN	1794
Safety railing	m	748

Table 4 shows the resultant costs of each solution for the studied vehicular bridges. It is evident that bridges made using structural steel beams are up to two times more expensive than those made with post-tensioned concrete. However, it is important to clarify that this comparison is related only to superstructure cost of each bridge (cost of beams, deck, and braces). Further study is required in near future for other cost generators as the substructure costs and environmental costs. Also, as explained previously, steel beams guarantee a better use of the space and a lower weight which could influence importantly the efficiency of the required substructure of the bridge.

Table 4. Costs comparison for the different types of bridges (millions COP).

Building material	Post-tensioned concrete beams				Structural steel beams			
	30 m	35 m	40 m	45 m	30 m	35 m	40 m	45 m
Concrete 28 MPa	231.15	319.28	434.08	530.22	51.16	59.69	68.22	76.74
Concrete 35 MPa	179.99	259.59	365.86	453.48				
Reinforcing steel 420 MPa	105.06	128.18	156.62	184.15	41.73	48.68	55.64	62.59
Prestressing Steel	295.42	433.26	598.68	794.09				
Steel A572-01 G-50					1362.47	1768.91	2190.72	2738.39
Steel profiles C 12×20.7					15.39	15.39	15.39	15.39
Steel profiles					18.21	21.21	24.21	27.21
Safety railing	44.88	52.36	59.84	67.32	44.88	52.36	59.84	67.32
Total cost	682.60	940.19	1257.33	1584.91	1533.84	1966.24	2414.01	2987.65

Figure 3 shows the cost trend in function of the span length for each one of the studied type of materials. It is evident that the bridge superstructure cost depends upon the span length; such dependence relation indicates that the longer the span the larger the cost. Cost of superstructures made using structural steel beams is about two times the cost of superstructures made with post-tensioned beams Such difference keeps almost constant through the range of spans evaluated.

Table 5 presents the trend equation for the cost of vehicular bridge superstructure made using post-tensioned concrete and structural steel beams with lengths varying between 30 and 45 m. The information of the Table 5 was calculated using Equation (3). The slope (coefficient of X) shows that cost for solutions made using steel beams impose a more rapid cost increase than those made using post-tensioned beams (96.184 > 60.482). It is evident that there is good fit of obtained equations ($R^2 \cong 1$).

Figure 3, Table 4, and Table 5 have been built considering that the two types of vehicular bridges designed with post-tensioned concrete and structural steel beams must have a similar functionality; such functionality has been defined as a function of the physical movement, *i.e.*, a target displacement has been guaranteed for each condition. In general, a bridge made using post-tensioned concrete beams could cost around the half of the cost of a bridge made using structural steel beams. In summary, That, the construction of post-tensioned concrete vehicular bridges could have greater economic viability than those designed with structural steel.

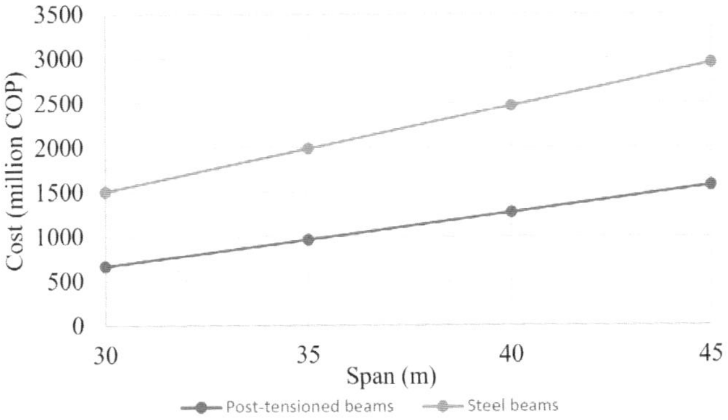

Figure 3. Cost comparison of vehicular bridge superstructure with spans ranging from 30 m to 45 m when designed using post-tensioned concrete or structural steel beams.

Table 5. Trend equations of the cost curves of short vehicular bridges ($30 \leq X \leq 45$).

Bridge girder material	Equation of the trend of the curve	R^2
Post-tensioned concrete	$Y = 60.482 * X - 1151.8$	0.9971
Structural steel	$Y = 96.184 * X - 1381.5$	0.9952

Y is the cost of the bridge superstructure in millions of Colombian pesos
X is the length of the bridge superstructure in meters ($30 \leq X \leq 45$)

4. Conclusions

The structural design of 4 bridges superstructures made using post-tensioned concrete beams and 4 bridges superstructures made using structural steel beams was carried out. Bridges superstructures designed with post-tensioned concrete girders resulted to be more economical than those designed with structural steel girders. In fact, solutions made of post-tensioned can cost around a half of what solutions made of steel beams cost. However, from the point of the physics (mechanics), the steel beams provide greater efficiency than post-tensioned beams because steel beams have a lower weight and can support a similar or larger demand than post-tensioned beams. Furthermore, mechanical efficiency of steel beams is around 1.5 times that of the post-tensioned beams for all the studied spans. Further analysis involving other important costs such as cost of substructure, social costs and environmental costs must be done before choosing any of the studied solutions.

References

[1] Muñoz Díaz E E 2011 *Ingeniería de Puentes. Reseña Histórica, Tipología, Diagnóstico y Recuperación* (Bogotá: Pontificia Universidad Javeriana)
[2] American Association of State Highway and Transportation Officials (AASTHO) 2017 *AASHTO LRFD Bridge Design Specifications* (Washington: American Association of State Highway and Transportation Officials)
[3] American Concrete Institute (ACI) 2014 *Building Code Requirements for Structural, ACI 318S-14* (United States of America: American Concrete Institute)
[4] Asociación Colombiana de Ingeniería Sísmica (AIS) 2014 *Norma Colombiana de Diseño de Puentes, LRFD –CCP 14* (Colombia: Ministerio de Ambiente, Vivienda y Desarrollo Territorial)
[5] Asociación Colombiana de Ingeniería Sísmica (AIS) 2010 *Reglamento Colombiano de Construcción Sismo Resistente, NSR10* (Bogotá: Ministerio de Ambiente, Vivienda y Desarrollo Territorial)

[6] El-Zohairy A, Alsharari F, Salimc H 2020 Analytical model and parametric study for externally post-tensioned reinforced concrete-steel composite beams *Structures* **27** 411
[7] Vasdravellis G, Tan B U E L, Kirkland B 2012 The effects of axial tension on the sagging-moment regions of composite beams *Journal of Constructional Steel Research* **72** 240
[8] California Department of Transportation (Caltrans) 2015 *Bridge Design Practice* (Sacramento: California Department of Transportation)
[9] Haas T 2014 Are reinforced concrete girder bridges more economical than structural steel girder bridges? A South African perspective *Jordan Journal of Civil Engineering* **8(1)** 43
[10] Delgado C A, Zúñiga B R 2015 *Comparación de Costos entre Puentes con Vigas de Acero, Concreto Reforzado y Postensado Considerando la Variación de la Luz Libre* (Cartagena de India: Universidad de Cartagena)
[11] Almeida J A, Armas A E 2019 *Diseño y Comparación Económica de la Superestructura de un Puente de 30 Metros de Luz Considerando Variantes de Losa con Vigas de Hormigón Postensado y Vigas-Losas-Cajón* (Quito: Escuela Politécnica Nacional)
[12] Montgomery D C, Runger G C1996 *Probabilidad y Estadística Aplicadas a la Ingeniería* (México: McGraw-Hill)

Evaluation of the fire resistance of steel-reinforced concrete-filled steel tubular columns with a circular cross-section

J P Rojas Suárez[1], J A Pabón León[1], and M S Orjuela Abril[1]
[1] Universidad Francisco de Paula Santander, San José de Cúcuta, Colombia

E-mail: sofiaorjuela@ufps.edu.co

Abstract. In the present investigation, an analysis of the fire resistance of the steel-reinforced concrete-filled steel tubular columns with circular cross-sections was carried out by means of numerical simulation. The development of the study was carried out by means of numerical simulation to predict the behavior of the column against fire. The results of the numerical model are validated by comparing the temperature levels obtained through experimental tests. From the results obtained, it is shown that the increase in the contact area between the steel and the concrete reduces the average temperature of the column, which implies a greater resistance to fire. The fire resistance of the columns with the steel profile designs are between 3.4 - 3.6 times higher compared to the column only made of concrete, which is an indication of the excellent performance of the steel-reinforced concrete-filled steel tubular columns with circular cross-sections columns. In general, the methodology proposed in this research allows the analysis of the thermal physical phenomena of the different columns used for the construction of buildings.

1. Introduction

Due to accelerated economic growth globally, there has been an increase in the construction of large-scale structures. The columns are the main elements used to support compression and bending loads. Since the importance of this element, it is necessary to guarantee the load capacity and ductility of the material. Among the existing column types, steel-concrete composite columns have high profitability and various advantages [1]. In this type of columns, there are the steel-reinforced concrete columns (SRC), concrete-filled steel tubular columns (CFST), concrete-filled double skin steel tubular columns (CFDST), steel-reinforced concrete-filled steel tubular columns (SRCFST), among others [2-5].

Steel reinforced concrete-filled steel tubular columns (SRCFST) are comprised of an outer steel tube, an embedded steel profile, and concrete. The above allows combining the advantages of the CFST and SRC columns. This type of column has been extensively investigated through experimental studies and numerical simulation to evaluate its behavior when subjected to static and dynamic loads [6-8]. The results obtained show that the steel profile embedded in the SRCFST columns significantly improves the load capacity, reduces the formation of cracks, presents greater seismic resistance, and improves the ductility of the column. Due to the above, SRCFST columns are widely used for building structures.

One of the main causes of massive disasters in buildings is the presence of fire [8]. Therefore, interest has been awakened to analyze the impact on the structure of the columns when they are subjected to high temperatures. Espinos, *et al.* [9] performed an evaluation of the fire resistance between CFST and CFDST columns by means of numerical simulation. Zhu, *et al.* [10] experimentally investigated the heating rate and fire resistance between SRCFST and CFST type columns. Tan, *et al.* [11] investigated the parameters of fire resistance, failure modes, internal force distribution and temperature distribution in CFST columns. Meng, *et al.* [12] experimentally studied SRCFST columns under different fire conditions, which include uniform fire and non-uniform fire.

Despite the various studies available in the literature, few studies have focused on SRCFST type columns under fire conditions. Due to the economic cost of creating experimental test facilities for the study of SRCFST columns and the difficulty of measuring variables such as deformation and temperature distribution, it is necessary to use other strategies to improve the understanding of the physical phenomena of SRCFST columns under the presence of fire.

The present research aims to study cross-section SRCFST columns that circulate under the presence of fire. The analysis is performed by finite element methods using OpenFOAM software [13]. The study involves the analysis of the temperature distribution variables and axial deformation under different types of steel profile designs and fire load ratios.

2. Methodology

This section describes the mathematical models, the properties of the materials and the computer-aided design (CAD) models of each type of column to be investigated.

The specific heat capacity of concrete is determined as a function of the temperature (T) range, as shown in Equation (1) [14].

$$\overline{\rho_c}\,\overline{s_c} = \begin{Bmatrix} \frac{19\rho_c s_c}{20} + \frac{\rho_w s_w}{20} & T \leq 100°C \\ \rho_c s_c & T > 100°C \end{Bmatrix}, \quad (1)$$

where ρ_w is the density of the moisture, $\overline{\rho_c}$ is the density considering the effect of water vapor, ρ_c is the density without considering the effect of water vapor, s_w is the specific heat capacity of the moisture, $\overline{s_c}$ is the specific heat capacity considering the effect of water vapor and s_c is the specific heat capacity without considering the effect of water vapor. The transitory thermal strain of concrete (ε_{th}) is determined by Equation (2) and Equation (3) [15].

$$\varepsilon_{th} = \kappa \cdot \varepsilon_e \frac{\sigma}{f_c} \quad T \leq 550\ °C, \quad (2)$$

$$\frac{\varepsilon_{th}}{\partial T} = -0.0001 \frac{\sigma}{f_c} \quad T > 550\ °C, \quad (3)$$

where ε_e is the thermal expansion strain of concrete, f_c is the cylinder strength of concrete, and κ is a constant of the model ($\kappa = 2$). To consider the effect of concrete softening due to tensile loads, the stress-deformation model proposed by Hon and Varma was used [16], as shown in Equation (4); where E_c is the concrete modulus of elasticity, ε is the strain, and σ is the stress, respectively.

$$\sigma = \begin{Bmatrix} E_c \cdot \varepsilon & \varepsilon < 0.09\frac{f_c}{E_c} \\ 0.099f_c - 0.1\varepsilon \cdot E_c & 0.09\frac{f_c}{E_c} \leq \varepsilon \leq 0.18\frac{f_c}{E_c} \\ 0.081f_c & \varepsilon > 0.18\frac{f_c}{E_c} \end{Bmatrix}, \quad (4)$$

The thermal properties for steel and concrete of the SRCFST column were based on the structural fire design rules EN 1992-1-2 [17] and EN 1993-1-2 [18], described by the European Committee for Standardization. A convection coefficient of 25 W/m^2K and an emissivity factor of 0.7 are established. The coefficient of thermal expansion for concrete was defined as $6 \times 10^{-6}\ K^{-1}$. The ambient temperature is set at a value of 27 °C. The emissivity of the fire was determined at a value of 1.

The behavior of the SRCFST columns in the presence of fire was a finite element analyzed with OpenFOAM software [13]. The geometric parameters defined for the construction of the SRCFST column were the external diameter, length of the column, the thickness of the steel tube, and support conditions. Additionally, the thermal and mechanical properties of the materials are established: concrete, steel profile, and steel tube. The load plates are modeled as perfectly elastic elements. The geometric shape of the components of the SRCFST columns are shown in Figure 1.

The development of the experimental tests was carried out in an electric oven. The length of the manufactured column was 1000 mm. For the validation of the results obtained by simulation, temperature sensors were installed inside the column.

(a)

(b)

(c)

Figure 1. Components of the SRCFST columns; (a) steel tube; (b) concrete; (c) steel profile.

3. Results

This section presents the results and discussions obtained from the investigation, which include the predictive capacity of the model and the analysis of the temperature distribution and axial deformation of the columns.

3.1. Experimental setup

To guarantee the reliability of the results predicted by the simulation, a comparison was made with the temperature obtained by the experimental tests in two locations of the column cross section: location A (outer surface of the column) and location B (geometric center of column). The results of the comparison are shown in Figure 2.

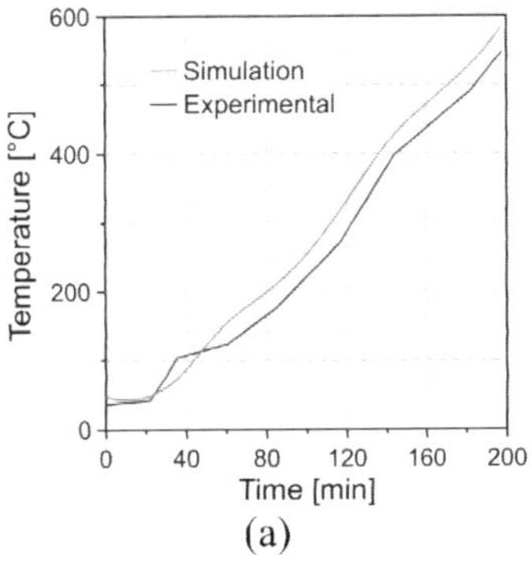

(a)
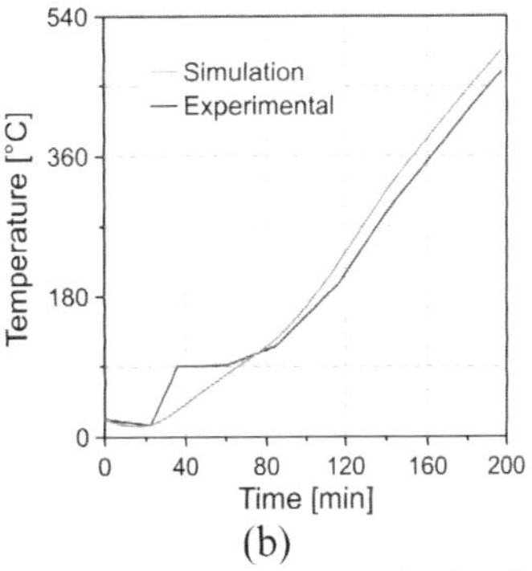

(b)

Figure 2. Comparison between the temperature obtained. experimentally and simulated; (a) location A; (b) location B.

From the results obtained in Figure 2 it was evidenced that the trend reported by the simulations describes the same behavior as the experimental measurements. An average relative error of 18% and 14% was observed for location A and location B, respectively. The largest deviation at location A is attributed to its proximity to the outer surface of the column, which may be exposed to uneven heat flow. Despite the above, the deviations obtained are considered adequate to study the physical behavior of SRCFST columns in the presence of fire.

3.2. Temperature distribution analysis

To analyze the effect of the embedded steel profile in SRCFST columns, different types of profile designs are analyzed. The types of simulated profile designs are shown in Figure 3. The results of the temperature analysis are described in Figure 4.

Figure 4 shows the temperature distribution of the cross-section of the three-column designs. In all designs, it was evidenced that the temperature decreases gradually from the outer surface to the geometric center of the column. In general, the presence of the steel profiles inside the column produces a greater temperature zone below 200 °C compared to the column formed only of concrete. Similar results are reported in the literature. This is attributed to the low thermal conductivity between steel and concrete. Comparing the designs of profiles type B and type C, it was evidenced that the latter presents a slightly lower temperature distribution compared to the design of profile type B. The above is a consequence of the greater contact area between the steel profile and the concrete. The previous results show that the zones of higher temperatures are considerably reduced when the steel profiles are embedded in the column, which is a significant benefit to improve fire resistance.

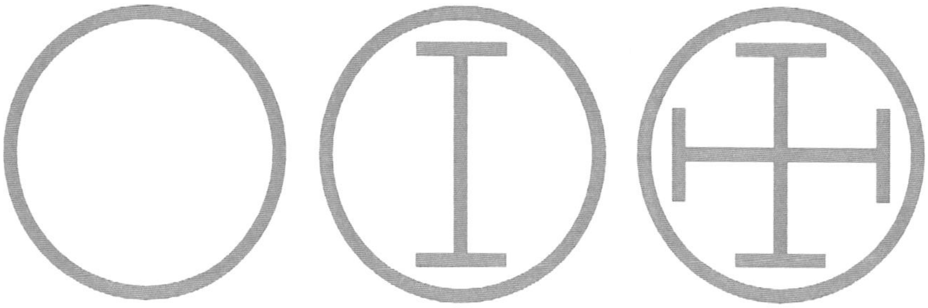

(a) (b) (c)

Figure 3. Column design; (a) type A; (b) type B; (c) type C.

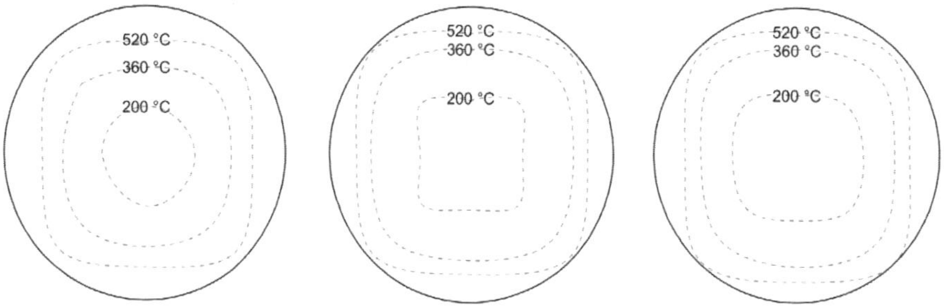

(a) (b) (c)

Figure 4. Temperature distribution for columns; (a) type A; (b) type B; (c) type C.

3.3. Analysis of axial deformation

Figure 5 shows the axial deformation of the column for the different types of columns (see Figure 3). In the case of type B and type C column designs, it was evidenced that these have a considerably higher fire resistance compared to columns only made of concrete (type A). This is demonstrated by considering the rapid drop in the deformation curve. This accelerated fall is a consequence of the deterioration of the rigidity of the column as it is in the presence of fire.

For type B and type C designs, it was observed that column failure occurs for an approximate time of 105 minutes and 111 minutes, respectively. This failure occurs as a consequence of the progressive buckling of the steel tube, which reduces the confining effect of the column. The higher fire resistance of type C column design compared to type B, is attributed to the regions of lower temperature in the cross-section of the column, as reported in the previous section. Additionally, the moment of inertia of the type C column is higher, which is another factor that contributes to the fire resistance of the column.

Figure 6 shows the axial deformation curve for different fire load ratios for the type C column. It was observed that the increase in the fire load ratio causes a considerable decrease in the fire resistance of the column. In the cases analyzed, a 40% and 60% decrease in catastrophic failure time was evidenced for a load ratio of 0.5 and 0.6 compared to a load ratio of 0.4, respectively. This is a consequence of the more accelerated loss of the stability and load capacity of the column.

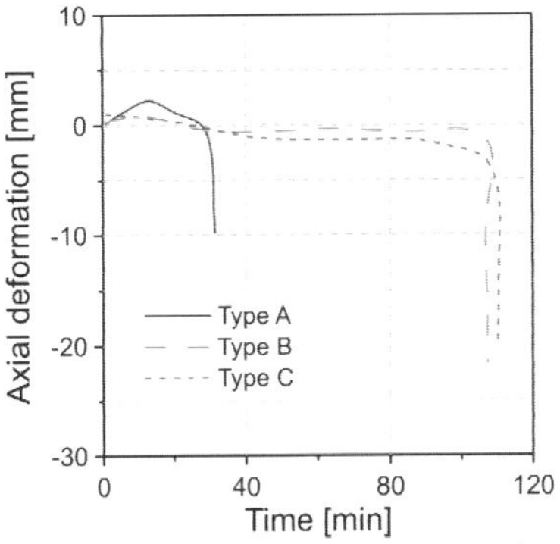

Figure 5. Axial deformation for different column designs.

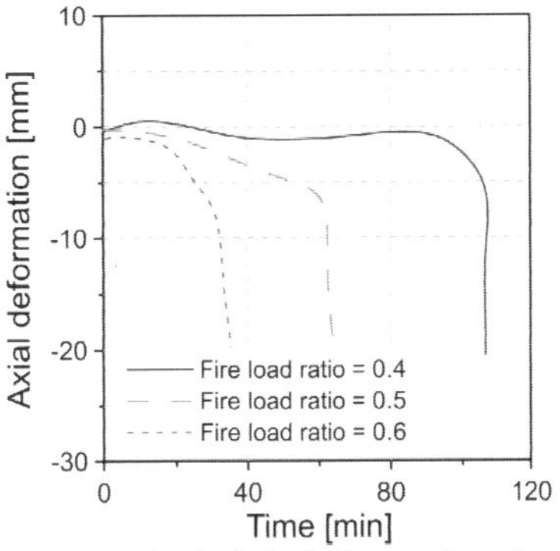

Figure 6. Axial deformation for different fire load ratios.

4. Conclusions

In the present investigation, an analysis was carried out by means of numerical simulation of the thermal behavior of the steel-reinforced concrete-filled steel tubular columns. The study involves the analysis of the temperature distribution variables and axial deformation under different types of steel profile designs and fire load ratios. The results obtained show that the thermal resistance between the steel profile and the concrete affects the temperature distribution in the cross-section of the column. In general, increasing the contact area between these two materials reduces the average column temperature, which implies greater resistance to fire.

The analysis of the axial deformation curve as a function of time shows that the embedding of the steel profiles in the columns considerably retards the accelerated failure. The fire resistance of the columns with the steel profile designs are between 3.4-3.6 times higher compared to the column only made of concrete, which is an indication of the excellent performance of the steel-reinforced concrete-filled steel tubular columns.

In general, the methodology proposed in this research allows the analysis of the thermal behavior of the different columns used for the construction of buildings. In this way, it contributes to the search and evaluation of new column designs, with the aim of improving their performance in the presence of fire.

References

[1] Kato B 1996 Column curves of steel-concrete composite members *Journal of Constructional Steel Research* **39(2)** 121

[2] Li W, Han L-H, Chan T-M 2014 Tensile behaviour of concrete-filled double-skin steel tubular members *Journal of Constructional Steel Research* **99** 35

[3] Zhou K, Han L-H 2018 Experimental performance of concrete-encased CFST columns subjected to full-range fire including heating and cooling *Engineering Structures* **165** 331

[4] Neuenschwander M, Knobloch M, Fontana M 2017 ISO standard fire tests of concrete-filled steel tube columns with solid steel core *Journal of Structural Engineering* **143(4)** 04016211

[5] He A, Liang Y, Zhao O 2020 Behaviour and residual compression resistances of circular high strength concrete-filled stainless steel tube (HCFSST) stub columns after exposure to fire *Engineering Structures* **203** 109897

[6] Shi Y-L, Xian W, Wang W-D, Li H-W 2020 Mechanical behaviour of circular steel-reinforced concrete-filled steel tubular members under pure bending loads *Structures* **25** 8
[7] Shi Y-L, Xian W, Wang W-D, Li H-W 2019 Experimental performance of circular concrete-filled steel tubular members with inner profiled steel under lateral shear load *Engineering Structures* **201** 109746
[8] Xian W, Wang W-D, Wang R, Chen W, Hao H 2020 Dynamic response of steel-reinforced concrete-filled circular steel tubular members under lateral impact loads *Thin-Walled Structures* **151** 106736
[9] Espinos A, Romero M L, Lam D 2016 Fire performance of innovative steel-concrete composite columns using high strength steels *Thin-Walled Structures* **106** 113
[10] Zhu M, Meng F, He B J 2016 Experimental research on fire resistance of steel tubular columns filled with steel reinforced concrete *Journal of Building Structures* **37(3)** 36
[11] Tan Q-H, Gardner L, Han L-H, Song T-Y 2019 Fire performance of steel reinforced concrete-filled stainless steel tubular (CFSST) columns with square cross-sections *Thin-Walled Structures* **143** 106197
[12] Meng F, Zhu M, Mou B, He B 2019 Residual strength of steel-reinforced concrete-filled square steel tubular (SRCFST) stub columns after exposure to ISO-834 standard fire *International Journal of Steel Structures* **19(3)** 850
[13] Greenshields C J 2015 *OpenFOAM The Open source CFD Toolbox Programmer's Guide Version 3.0.1* (London: OpenCFD Ltd)
[14] Lie T T 1994 Fire resistance of circular steel columns filled with bar-reinforced concrete *Journal of Structural Engineering* **120(5)** 1489
[15] Anderberg G Y, Thelandersson S 1976 *Stress and Deformation of Concrete at High Temperatures. 2 Experimental Investigation and Material Behaviour Model* (Lund: Lund University)
[16] Hong S, Varma A H 2009 Analytical modeling of the standard fire behavior of loaded CFT columns *Journal of Constructional Steel Research* **65(1)** 54
[17] Comité Européen de Normalisation (CEN) 2002 *Eurocode 1: Actions on Structures-Part 1-2: General Actions-Actions on Structures Exposed to Fire, EN 1991-1-2* (Belgium: Comité Européen de Normalisation)
[18] Comité Européen de Normalisation (CEN) 2005 *Eurocode 3: Design of Steel Structures-Part 1-2: General Rules-Structural Fire Design, EN 1993-1-2* (Belgium: Comité Européen de Normalisation)

Journal of Physics: Conference Series **2153** (2022) 012006 doi:10.1088/1742-6596/2153/1/012006

Resistance under Marshall monotonic load for asphalt concrete mixtures

Y W Yung-Vargas[1], A Rodríguez-Lizcano[1], and C A Peña-Soto[1]
[1] Grupo de Investigación en Infraestructura Vial, Universidad Francisco de Paula Santander, San José de Cúcuta, Colombia

E-mail: yeewanyv@ufps.edu.co, adrianarodriguez@ufps.edu.co

Abstract. The dense hot MDC-19 type asphalt mixes are considered, by the "Instituto Nacional de Vías" in Colombia, as continuous grading asphalt mixes (asphalt concrete). These constitute most of the surface course, in the structures of in-service pavements, being the object of study and research in different projects to ensure their durability. In the present investigation, unlike other investigations, the mechanical behavior under Marshall monotonic load was studied in the laboratory between MDC-19 dense type asphalt mixtures, comparing plant-produced and laboratory-produced asphalt mixtures. To carry out this process, samples of uncompacted asphalt mixtures were taken, produced in four fixed plants, with which Marshall-type briquettes were compacted. Likewise, samples of mineral aggregates and asphalt cement were obtained from the same plants, which constitute the mixtures raw material produced there. With these materials, briquettes with the same characteristics were mixed and compacted. Subsequently, the resistance under Marshall monotonic load was determined on the briquettes manufactured in plant and laboratory. The optimal asphalt cement content was compared between plant and laboratory-produced mixtures. An increase in Marshall Stability was found in the briquettes made with plant-produced mixtures, while these required a greater amount of asphalt cement for their production.

1. Introduction

Dense hot mixes are known in Colombia as asphalt concrete mixes following the standard of "Instituto Nacional de Vías (INVÍAS)" [1]. These are mixtures that are placed as layers of asphalt bases and surface course in flexible pavements, the latter being the most used worldwide (more than 90%) according to "Asociación de Productores y Pavimentadores Asfálticos de Colombia (ASOPAC)" [2], and European Asphalt Pavement (EAPA) [3,4]. These mixtures are characterized by having a percentage of air voids between 3% and 9%, which makes them waterproof, also presenting high resistance to traffic loads and weathering agents.

They are manufactured in fixed or road plants at a high temperature (depending on the viscosity of the asphalt binder to be used) and are made up of the union between asphalt cement and mineral aggregates that vary in size from passing 1" sieve to passing sieve No. 200 [1]. Among the factors that affect the behavior of asphalt concrete mixtures are the source of the aggregates, their distribution, size and shape, type of asphalt cement and its rheological characteristics, mixing temperature, compaction temperature, and mixing process in the plant and the laboratory. Authors who have studied some of these factors include [5-8].

In the present investigation, the study of asphalt concrete mixtures was deepened, considering the influence that the source of the mineral materials has on the resistance, under a monotonic load, of MDC-19 type mixtures, manufactured in the plant and the laboratory.

2. Materials and methods

Samples of MDC-19 type asphalt mixtures (loose, not compact) were obtained from four production plants (P1, P2, P3, and P4). These mixtures were subjected to a compaction process in the laboratory, to obtain 4" in diameter and 2.5" thick briquettes. The traffic was considered greater than 5.0 equivalent axles of 80 KN in the design lane, for INVÍAS [1], which required compaction at 75 strokes on each face [9]. Likewise, samples of mineral aggregate and asphalt cement type CA 60-70 mm/10 were obtained, with which the asphalt mixtures are produced in each plant. Plant P1 and plant P4 obtain the aggregate from the same source (F1), and plant P2 and plant P3 obtain them from sources F2 and F3 respectively.

The asphalt cement for the four plants comes from the same source. The mineral materials and asphalt cement type CA 60-70 mm/10 were evaluated in the laboratory considering the characteristic tests recommended by [1]. Subsequently, with the aggregates and asphalt cement from each plant, we proceeded to design and manufacture MDC-19 asphalt concrete mixtures in the laboratory, mixing and compacting 4" diameter, and 2.5" thick briquettes, at mixing and compaction temperatures referred to by the viscosity of the asphalt cement used. Table 1 shows the granulometry for the manufacture of an MDC-19. Each compacted briquette had an approximate weight of 1200 g.

Three briquettes were manufactured for each percentage of asphalt cement considered (4.5%, 5.0%, 5.5%, 6.0%, 6.5%). Both the compacted briquettes with the loose mixtures from each plant, as well as the laboratory-compacted briquettes, were subjected to the Marshall test to determine their resistance under monotonic load. This allowed us to calculate the optimum percentage of asphalt in the mixtures and the stability/flow ratio, known as Marshall stiffness. In Colombia, this test is regulated by the INVÍAS standard [1].

A more detailed description of this procedure can be found in American Association of State Highway and Transportation (AASHTO T 245) [10], asphalt mix design methods [11,12], American Society for Testing and Materials (ASTM D6926-10) [13], and "Una Norma Española (UNE-EN 12697-34)" [14]. The asphalt content extraction test [1] and aggregate granulometry test [1] were carried out on the loose asphalt mixtures from each plant, to determine the optimum percentage of asphalt.

Table 1. Granulometric distribution for an MDC-19 mixture.

Sieve	Sieve (mm)	Percent passing (%)
3/4"	19.000	100.0
1/2"	12.500	87.5
3/8"	9.500	79.0
No. 4	4.750	57.0
No. 10	2.000	37.0
No. 40	0.430	19.5
No. 80	0.180	12.5
No. 200	0.075	6.0

3. Results

Table 2 shows the results for the aggregate characterization of the four plants evaluated; variability is observed in the parameters for hardness, particle shape, and material cleanliness, even though two samples came from the same source. The foregoing is due to the process and site of exploitation of the materials. The tests were based on the INVÍAS standard [1]. Table 3 shows the results of the asphalt cement used in each plant to produce the asphalt mixtures, which corresponds to the same asphalt cement used for the manufacture of the briquettes in the laboratory. It can be noted that the asphalt cement CA 60-70 mm/10, complies with the parameters´ ranges proposed by the INVÍAS standard [1].

Table 2. Results of the characterization of the aggregate of the four plants.

Test	Recommended	P1	P2	P3	P4
Specific gravity/fine aggregate absorption	-	2.55/1.03	2.55/1.70	2.55/1.60	2.54/1.62
Specific gravity/coarse aggregate absorption	-	2.57	-	-	-
Fractured particles (1 side)	85% minimum	90.92	91.07	93.31	98.76
Flattening index	10% maximum	3.01	3.47	13.33	11.90
Elongation index	10% maximum	7.68	14.07	10.73	8.82
Plasticity index	Not plastic	NP	NP	NP	NP
Sand equivalent (%)	50 minimum	56.82	52.08	33.33	43.28
Abrasion in Los Angeles machine	25% maximum	34.90	24.67	36.92	37.03

Table 3. Results of the characterization of asphalt cement CA 60-70 mm/10.

Test	Unit	Recommended value	P1	P2	P3	P4
Specific gravity (Kg/m^3)	-	-	1011	990	1011	1010
Penetration (25°C-100 g-5 s)	mm	6-7	61	63	63	65
Penetration Index	-	-1.2/+0.6	-0.79	-0.90	-0.65	-0.80
Viscosity (60°C)	Poise	1500 minimum	2340	2120	2120	2140
Softening point	°C	48-54	49.8	48.9	50.0	49.2
Ductility (25°C-5cm/min)	cm	100 minimum	136.5	140.0	130.0	145.0
Flash point	°C	230 minimum	298	296	304	278

Table 4 and Table 5 show, respectively, the mean of the granulometry and the mean of the percentage of asphalt cement obtained from the characterization of the asphalt mixtures, in a loose state, produced in the four plants. Variation can be noted in the percentages of material that pass the sieves with the highest denomination, and an equal trend in the sieves that pass No. 4 sieve. An average value of the asphalt cement is observed oscillating around 5.37%.

Table 4. Average granulometry in asphalt mixtures in plants.

Sieve	Percent passing (%) [1]	P1	P2	P3	P4
¾"	100	100.00	100.00	100.00	100.00
½"	80-95	80.01	82.23	90.42	90.31
3/8"	70-88	65.21	72.00	80.84	78.81
No. 4	49-65	48.19	44.50	53.94	54.32
No. 10	29-45	37.06	32.74	35.96	39.21
No. 40	14-25	24.91	17.62	21.91	26.22
No. 80	8-17	18.13	10.14	13.14	16.75
No. 200	4-8	11.45	6.28	7.10	8.91

Table 5. Average asphalt cement in asphalt mixtures in plants.

Plant	P1	P2	P3	P4
Asphalt percent (%)	5.17	5.48	5.51	5.33

Table 6 shows the results of the stability/flow test (S/F) carried out on the compacted briquettes with the mixtures from each plant and on the briquettes produced in the laboratory with the materials (mineral aggregates and asphalt cement) from the same plants. Compliance with the parameters required by the regulations is observed.

Table 6. Stability/flow test results.

Parameters	P1		P2		P3		P4	
	MP	ML	MP	ML	MP	ML	MP	ML
Stability S (N)	13750	13100	16234	13700	16357	12000	18835	13400
Flow F (mm)	3.10	2.98	3.50	2.80	3.40	2.60	3.52	3.05
S/F (KN/mm)	4.44	4.40	4.64	4.89	4.81	4.62	5.35	4.39

MP is mix in plant, and ML is mixing in laboratory.

Table 7 shows the percentages of asphalt cement found in the extraction test carried out on the mixtures made in the plant, and the percentages of asphalt cement obtained by the stability/flow test and analysis of gravimetric and volumetric relationships carried out on the mixtures produced in the plant and those made in the laboratory. It is noteworthy that there is a greater requirement for asphalt cement in the mixtures manufactured in the plant, with respect to the mixtures prepared in the laboratory.

The above, because the aggregates in the laboratory were separated into each of the sizes indicated for an MDC-19 (From pass 3/4" sieve to pass No. 80 sieve), and subsequently washed and dried; contrary to the process applied in the plant, where the materials are separated into fractions (coarse, medium and fine), and these are not washed and dried. This makes the finer materials adhere to the thick ones, requiring more amount of bitumen. The asphalt cement content obtained by the extraction process on the mixtures produced in the plants is lower than the optimal percentages obtained in the Marshall design.

Table 7. Percentages of asphalt cement CA.

Plant	P1	P2	P3	P4
CA% by extraction	5.17	5.48	5.51	5.33
CA% Plant-produced mix	5.50	5.88	5.75	5.70
CA% mixtures produced in the laboratory with materials from each plant	5.20	5.40	5.20	5.50

4. Conclusions

The Marshall test on the mixtures produced in the plant and the laboratory determines that the aggregate preparation process, as well as the source of the material, influences the determination of the optimum percentage of asphalt cement in asphalt mixtures; the present study ranged around 5.71% for mixtures produced in the plant, and around 5.32% for mixtures produced in the laboratory, which represents 0.4% of additional consumption in the plant. Greater consumption of asphalt is required in the plant due to the higher content of fines in the mix, pass No. 200 sieve.

Marshall stability is above the minimum estimated by the standard of "Instituto Nacional de Vías", with higher values found in the plant. The higher the content of fines, the higher the consumption of asphalt cement, and the higher the resistance under Marshall monotonic load. It is recommended to adjust to the asphalt mix design in the plant, considering that the aggregates are not separated, nor are they washed before mixing them. In future research it is recommended to determine on asphalt mixtures manufactured in plant and laboratory, the resistance under cyclic load and the resistance to permanent deformation.

References

[1] Instituto Nacional de Vías (INVÍAS) 2012 *Especificaciones Generales de Construcción de Carreteras* (Colombia: Instituto Nacional de Vías)

[2] Asociación de Productores y Pavimentadores Asfálticos de Colombia (ASOPAC) 2004 *Cartilla del Pavimento Asfáltico* (Colombia: Asociación de Productores y Pavimentadores Asfálticos de Colombia)

[3] European Asphalt Pavement Association (EAPA) 2001 *Asphalt in Figures, Key Figures of the European Asphalt Industry* (Brussels: European Asphalt Pavement Association)

[4] European Asphalt Pavement Association (EAPA) 2015 *Asphalt in Figures, Key Figures of the European Asphalt Industry* (Brussels: European Asphalt Pavement Association)

[5] Leandro R P, Vasconcelos K L, Bernucci L L B 2017 Evaluation of the laboratory compaction method on the air voids and the mechanical behavior of hot mix asphalt *Construction and Building Materials* **156** 424

[6] Sousa J B, Harvey J, Painter L, Deacon J A, Monismith C L 1991 *Evaluation of Laboratory Procedures for Compacting Asphalt-Aggregate Mixtures* (Washington: Strategic Highway Research Program)

[7] Kennedy T W 1977 Characterization of asphalt pavement materials using the indirect tensile test, *Association of Asphalt Paving Technologists Procceedings* **46** 132

[8] National Cooperative Highway Research Program (NCHRP) 2011 *A Manual for Design of Hot-Mix Asphalt with Commentary, Report 673* (Washington: Transportation Research Board)

[9] Asphalt Institute 2001 *Superpave Mix Design* (Lexington: Asphalt Institute)
[10] American Association of State Highway and Transportation Officials (AASHTO) 2015 *Standard Method of Tets for Resistance to Plastic Flow of Asphalt Mixtures Using Marshall Apparatus, AASHTO T245* (United States of America: American Association of State Highway and Transportation Officials)
[11] Asphalt Institute 2014 *Aphalt Mix Design Methods* (Lexington: Asphalt Institute)
[12] Garnica P, Delgado H, Gómez J A, Romero S A, Alarcón H A 2004 *Aspectos del Diseño Volumétrico de Mezclas Asfálticas* (Sanfandila: Instituto Mexicano del Transporte)
[13] American Society for Testing and Materials (ASTM) 2010 *Standard Practice for Preparation of Bituminous Specimens Using Marshall Apparatus, ASTM D6926-10* (United States of America: American Society for Testing and Materials)
[14] Asociación Española de Normalización y Certificación (AENOR) 2013 *Mezclas Bituminosas. Métodos de Ensayo para Mezclas Bituminosas en Caliente. Parte 34: Ensayo Marshall, UNE-EN 12697-34* (España: Asociación Española de Normalización y Certificación)

Evaluation of recycled asphalt pavement in Colombia

Y W Yung-Vargas[1], H A Rondón-Quintana[2], and J E Córdoba-Maquilón[3]
[1] Grupo de Investigación en Infraestructura Vial, Universidad Francisco de Paula Santander, San José de Cúcuta, Colombia
[2] Centro de Estudios en Pavimentos y Materiales Sostenibles, Universidad Distrital Francisco José de Caldas, Bogotá, Colombia
[3] Grupo de Investigación Vías y Transporte, Universidad Nacional de Colombia, Medellín, Colombia

E-mail: yeewanyv@ufps.edu.co, harondonq@udistrital.edu.co

Abstract. Reclaimed asphalt pavements are obtained from existing pavements through recovery techniques and are used for new asphalt mixtures production with the inclusion of virgin aggregates and asphalt cement, constituting environmentally friendly mixtures at a lower total cost, by requiring fewer quantities of new materials. This research, unlike the studies found on the subject, focuses on the study of reclaimed asphalt pavement by analyzing its granulometric distribution and asphalt cement content. For this purpose, representative reclaimed asphalt pavement samples were taken from four Colombia cities, to verify their heterogeneity. The obtained reclaimed asphalt pavement was analyzed in the laboratory, to obtain the material granulometric distribution, through extraction and asphalt content tests, following Colombian regulations from the "Instituto Nacional de Vías". The results show that the granulometry of all cities has a central tendency (average) that moves towards an upper limit in the 2 mm particles. When analyzing the granulometry separately of the four cities, it is observed that they present a different trend in their granulometry, which shows their heterogeneity. The asphalt cement content presented values between 4.0% and 5.0%.

1. Introduction

Millions of tons of reclaimed asphalt pavements (RAP), a product of the milling of the roadways, are discarded in the world, product of the milling of the roadways that contain it. A fraction, of these milled materials, is reused for the elaboration of new mixtures, either for later use of asphalt layers or as stabilized bases. The remaining material is taken to landfills which generate a strong environmental impact. The RAP that is milled varies in its size and composition depending, among other factors, on the aggregates source (quarry, river, or combination of both), granulometric composition (material size and distribution), asphalt content, characteristic of asphalt, degree of aging of the mixture (caused by external agents such as climatic conditions during service years), and traffic, making RAP a heterogeneous material.

The procedure and equipment used in the roadways for the extraction of RAP also influence its properties. In the milling process, the size of the aggregates is reduced due to the mechanical degradation to which they are subjected, being able to obtain grinding, crushed, or in large pieces. Han J, *et al.* [1] took milled RAP from three projects, separating aggregate from asphalt cement through ignition and centrifugation processes. They found that the aggregates obtained, by both methods, were reduced in size compared to the size of the aggregates used in the initial mix, reporting a 4% increase in the fine particles that passed the 75-micron sieve. On the other hand, Solaimanian and Tahmoressi [2] took different sizes of RAP and determined the influence that the variation of these sizes has within a new mixture while using 35%, 40%, and 50% RAP fractions. They found that using higher percentages of this material increased the variation in the new mixture gradation; They also reported that when using RAP from different sources, the new mixtures void content increased by increasing the RAP percentage used.

Other investigations carried out on RAP and on new mixtures produced with it, focus on the study of the new mixtures mechanical resistance and durability parameters, with different RAP contents, which vary from 10% to 100%, such as [3-10]. In addition, the mechanical behavior of recycled mixtures has been studied regarding the content of voids and Marshall stiffness [11-13]. In the same way, researchers such as Lee, *et al.* [14] and Silva, *et al.* [15] have reported studies regarding permanent deformation in recycled mixtures, and fatigue Izaks, *et al.* [16]. The few studies on the RAP granulometric composition and its influence on the behavior of mixtures generate uncertainty in the determination of the required percentages of RAP and new components (stone aggregate and asphalt cement) in the manufacture of hot mix asphalt (HMA).

Therefore, this research, unlike others, aims to deepen the study of RAP mixtures, demonstrating the heterogeneity in the composition of the materials that compose them. This is the first stage, and fundamental basis, for an investigation that is carried out on the influence that this variability has on the manufacture of new HMA.

2. Materials and methods

For the present research, to verify the heterogeneity of the asphalt mixtures, a total of 31 RAP samples were taken from 4 cities in Colombia. Asphalt content extraction test (INV E-732-13) [17] and aggregate granulometry test (INV E-782-13) [18] were carried out on each sample, to determine the optimum asphalt content and the granulometry of the mixtures.

To perform these tests, approximately 1200 grams of RAP material were taken per test, subjecting it to drying and subsequent centrifugation process, for the extraction of asphalt, as referred to in the standard INV E-732-13 [17]. After asphalt extraction from the mixture, the aggregates were taken to the oven, weighed, and washed on the No. 200 sieve to be dried again and subjected to sieving (INV E-782-13) [18]. Once the previous process was finished, the characteristic granulometric curve of each mixture was determined.

The process was repeated for all the materials from each city. Subsequently, the mean of the granulometry and the percentage of asphalt cement that make up the RAP was determined, for the evaluated cities of Colombia. Through a statistical and variance analysis (unifactorial or one-way analysis of variance (ANOVA)), the optimal asphalt content of the four cities and the most representative granulometry were determined.

3. Results and discussion

After performing the tests to extract the asphalt content in all the samples, different granulometries were found for the four cities, verifying that they are heterogeneous. Taking the results for each of the granulometries, the mean was determined for all the cities whose results are shown in Table 1.

The aggregates show a trend towards the upper limit of the INVÍAS [18] regulation for an MDC-19 asphalt mix, which means that it tends to be a fine material. This is related to the milling process by which the RAP samples were obtained. Milling affects the particle size of the aggregates, with the consequent increase in fine particles content. Table 2 shows the dispersion parameters of the four Colombia cities analyzed.

With the results of the mean values of each size in the granulometries of the four cities, an ANOVA was performed. Table 3 shows the results of F (statistic), F critical, and significance p-value. F values greater than F critical, and significance values less than 0.05 are observed; this indicates a greater statistical significance and allows us to conclude that there is heterogeneity in the aggregate particle size compared among the four cities analyzed.

Table 1. Results of the mean in the granulometry of all the cities.

Sieve	Sieve (mm)	Average	Inferior limit [18]	Superior limit [18]
3/4"	19.000	98.0	100	100
1/2"	12.700	89.4	80	95
3/8"	9.520	80.5	70	88
N°4	4.750	61.6	49	65
N°10	2.000	45.9	29	45
N°40	0.425	27.8	14	25
N°80	0.177	18.3	8	17
N°200	0.074	9.2	4	8

Table 2. Granulometric dispersion parameters for the analyzed cities.

Passing sieve	Dispersion measurements of the different sizes of particles							
	3/4"	1/2"	3/8"	N°4	N°10	N°40	N°80	N°200
Average	98.03	89.42	80.54	61.58	45.94	27.81	18.27	9.20
Variance (S^2)	2.46	22.54	25.48	24.11	13.34	7.60	3.00	1.20
Standard deviation (SD)	1.57	4.75	5.05	4.91	3.65	2.76	1.73	1.09
Coefficient of variation (cv)	1.60	5.31	6.27	7.97	7.95	9.91	9.48	11.89

Table 3. Results of the analysis of variance (ANOVA) in the granulometries of the four cities.

Passing sieve	No. ¾"	No. ½"	No. 3/8"	No. 4	No. 10	No. 40	No. 80	No. 200
F	12.35	4.578	7.797	13.043	24.465	44.237	51.927	17.397
F critical	2.96035							
p-value	<0.05							

The asphalt cement was extracted from the asphalt mixtures and the asphalt content with respect to the total mixture was determined. Based on the asphalt content extracted in the RAP, calculation of the mean and the dispersion parameters (variance, standard deviation, and coefficient of variation) was performed. Table 4 shows that the mean asphalt cement content oscillates around 4.61%, with a standard deviation of 0.61.

Subsequently, an ANOVA was carried out on the results of the asphalt extractions of the RAP of the roadways for the four cities studied. The results of the ANOVA show an $F = 12.055$ greater than the F (critical) = 2.975, and a significance of $p\text{-value} < 0.05$. A significance below 0.05 is observed, which represents greater statistical significance, indicating variation in the means of the asphalt cement values obtained in the four cities.

Table 4. Dispersion parameters in the asphalt content.

Dispersion measure	Asphalt content (%)
Average	4.61
Variance (S^2)	0.37
Standard deviation (SD)	0.61
Coefficient de variation (cv)	13.16

4. Conclusions

The tests of "Instituto Nacional de Vías" performed on the asphalt extraction process and granulometry of the aggregate particles reports that the mean of the granulometry extracted from the four evaluated cities shows a central tendency within the granulometric band for an MDC-19 mixture in the sizes of the aggregates that pass the No. ¾" sieve to No. 10 sieve. On the contrary, for the aggregates that pass from No. 10 sieve to No. 200 sieve, the trend is above the upper limit of the granulometric range for an MDC-19 mixture, indicating a higher fine fraction in the reclaimed asphalt pavements. This higher fraction is due to the mechanical wear to which the material has been subjected when it is extracted from the roadway, which is variable depending on the methodology used in the field.

The analysis of variance showed significant differences in the mean values of the size of the aggregate particles extracted from the reclaimed asphalt pavements in the four cities, demonstrating that it is a heterogeneous material. The heterogeneity is also reflected in the percentages of asphalt cement extracted from the reclaimed asphalt pavements of the cities evaluated, some above the average and others below it. The mean asphalt cement content, in the reclaimed asphalt pavements mixes evaluated, is on the order of 4.6%. The heterogeneity in the composition of the aggregates and the asphalt cement content are parameters to consider when designing a new asphalt mix that includes reclaimed asphalt pavements particles.

References

[1] Han J, Thakur S C, Chong O, Parsons R L 2011 *Laboratory Evaluation of Characteristics of Recycled Asphalt Pavement in Kansas, Report KU-09-2* (Lawrence: University of Kansas)
[2] Solaimanian M, Tahmoressi M 1996 Variability analysis of hot-mix asphalt concrete containing high percentage of reclaimed asphalt pavement *Transportation Research Record: Journal of the Transportation Research Board* **1543(1)** 89
[3] Salaimanian M, Kennedy T W 1995 *Production Variability Analysis of Hot-Mixed Asphalt Concrete Containing Reclaimed Asphalt Pavement* (Austin: University of Texas)
[4] Ma T, Wang H, Huang X, Wang Z, Xiao F 2015 Laboratory performance characteristics of high modulus asphalt mixture with high-content RAP *Construction and Building Materials* **101** 975
[5] Liphardt A, Król J, Radziszewski P 2016 Influence of polymer modified binder content from RAP on stone mastic asphalt rutting resistance *Procedia Engineering* **153** 407
[6] Riccardi C, Cannone-Falchetto A, Losa M, Wistuba M 2016 Back-calculation method for determining the maximum RAP content in Stone Matrix Asphalt mixtures with good fatigue performance based on asphalt mortar tests *Construction and Building Materiales* **118** 364
[7] Avilés Lorenzo J 2002 *Estudio de la Tenacidad de los Microaglomerados Reciclados en Caliente Mediante el Ensayo BTD, Efecto del Tipo y Contenido de Betún* (Barcelona: Universitat Politécnica de Catalunya)
[8] Hajj E Y, Sebaaly P E, Kandiah P 2010 Evaluation of the use of reclaimed asphalt pavement in airfield HMA pavements *Journal of Transportation Engineering* **136(3)** 181
[9] Valdés G, Pérez-Jiménez F, Miró R, Martínez A, Botella R 2010 Experimental study of recycled asphalt mixtures with high percentages of reclaimed asphalt pavement (RAP) *Construction and Building Materials* **25(3)** 1289
[10] Pradyumna T A, Mittal A, Jain P K 2013 Characterization of reclaimed asphalt pavement (RAP) for use in bituminous road construction *Procedia Social and Behavioral Sciences* **104** 1149
[11] Dos Santos S, Partl M N, Poulikakos L D 2015 From virgin to recycled bitumen A microstructural view *Composites Part B: Engineering* **80** 177
[12] Noferini L, Simone A, Sangiorgi C, Mazzotta F 2017 Investigation on performances of asphalt mixtures made with reclaimed asphalt pavement: effects of interaction between virgin and RAP bitumen *International Journal of Pavement Research and Technology* **10(4)** 322
[13] Gómez-Meijide B, Ajam H, Lastra-González P, Garcia A 2018 Effect of ageing and RAP content on the induction healing properties of asphalt mixtures *Construction and Building Materials* **179** 468
[14] Lee K W, Soupharath N, Shukla A, Franco C A, Manning F J 1999 Rheological and mechanical properties of blended asphalts containing recycled asphalt pavement binders *Journal of the Association of Asphalt Paving Technologists* **68** 89
[15] Silva H M R D, Oliveira J R M, Jesus C M G 2012 Are totally recycled hot mix asphalts a sustainable alternative for road paving? *Resources, Conservation and Recycling* **60** 38
[16] Izaks R, Haritonovs V, Klasa I, Zaumanis M 2015 Hot mix asphalt with high RAP content *Procedia Engineering* **114** 676
[17] Instituto Nacional de Vías (INVÍAS) 2013 *Extracción Cuantitativa del Asfalto en Mezclas para Pavimentos, INV E-732-13* (Colombia: Instituto Nacional de Vías)
[18] Instituto Nacional de Vías (INVÍAS) 2013 *Análisis Granulométrico de los Agregados Extraídos de Mezclas Asfálticas, INV E-782-13* (Colombia: Instituto Nacional de Vías)

Corrosion resistance evaluation of carbon and vanadium-based sputtered coatings on steel substrates

C Vera[1], R Sorzano[1], L Ardila[1], G Orozco[1], and W Aperador[1]
[1] Universidad ECCI, Bogotá, Colombia

E-mail: gorozcoh@ecci.edu.co

Abstract. Steels are in constant contact with fluids that could generate corrosion regardless the application in which this steel is located. AISI-SAE 1045 like steels is widely used in different applications in engineering, even several of these parts made of this steel suffers wear processes. The synergy between corrosion and wear phenomena exacerbates the detriment of some physical properties of the material conducing it to a failure. A potential alternative to avoid this issue is to coat the material surface with an anticorrosive material, and among different techniques to produce coatings, physical vapor deposition ones are environmentally friendly, secure and with excellent properties on the final product. We report the production of coatings based on vanadium and carbon on AISI-SAE 1045 steels substrates varying some of the deposition parameters in a sputtering coatings machine. A 2^3-factorial design of experiments was done with power applied to the vanadium target, power applied to the carbon target and temperature as active factors with two levels each one. A relevant effect of the power applied to V target and temperature on the anticorrosive properties of the coatings was found, thus increasing these factors levels always gives higher surface roughness and higher corrosion rates, this result together provides an important insight into the values that must be considered to achieve good anticorrosive properties on the material. Overall, these results indicate that with low V target power and room temperature, and high C target power the lowest corrosion rates and roughness of the group are achieved, both results agree.

1. Introduction

The industrial uses of steels are worldwide spread and plays an important role in the modern commodities. Among the applications of steels are molds for polymer shapes, sprockets, wedges, cranks, chains, low speed gears, machinery parts, agricultural machinery parts, agricultural tools, and other industrial parts [1]. The industry of pieces that undergo wearing and fatigue process has historically had different problems in terms of wear and corrosion [2]. The synergy of this phenomena is one of the most widely responsible of a reduction in the surface properties of the pieces, making this issue a major area of interest within the field of materials science.

AISI-SAE 1045 steel is increasingly recognized as an important one due to its special features as well as the practical applications in which is involved. One of this application is the production of molds for stretch blow molding and the production of polyethylene terephthalate (PET) bottles [3]; in this industry the pieces are in constant contact with fluids making them more suitable for corrosion processes. This aggressive medium leads to an anodic dissolution of the steel surface acting as a galvanic corrosion system [4].

Previous studies have reported the use of different coatings methods to enhance the anticorrosive behavior of pieces made of AISI-SAE 1045 steel: thermo-reactive diffusion, laser cladding and direct current (DC) reactive magnetron sputtering; one of the compounds classically intended for this application of coating on this steel is based on carbon and vanadium due to the uncommon combination of properties like hardness, thermal conductivity, grain size and high coefficient of friction [5,6]. However, even when these methods are excellent options of coatings production questions have been raised about the use of a detailed design of experiments to evaluate the effect of the production parameters in its properties.

Physical vapor deposition (PVD) techniques to coat substrate materials are a group of techniques to produce coatings on different kind of materials which along with the benefits of vacuum produce highly homogeneous coatings with low roughness and high density [7]. To examine the significance of the parameters used in the deposition of carbon and vanadium-based films on the anticorrosive features of AISI-SAE 1045 steel substrates, this research investigates several ranges of sputtering power applied to the targets and substrate temperatures, this research will stablish the basis from a technical point of view of the deposition conditions like vanadium target power, carbon target power and temperature of deposition and its relevance on the corrosion behavior of coated AISI-SAE 1045 steel that is widely used in the engineering industry.

2. Experimental details

It was considered that quantitative measures would usefully supplement and extend the scopes of this paper.

2.1. Coating's deposition

In this research the deposition parameters are of special interest. CY-VTC-600-3HD three-target magnetron sputtering coater (Zhengzhou C Y, Scientific Instruments Co., Ltd) was used in DC mode. Factors as substrate temperature, power, current and voltage applied to the target, samples rotation, gases flow rate and substrate to target distance can be controlled. Three factors were studied in a 2^3 full factorial design with three replicates and factor levels set as shown in Table 1; DC power applied to 99.9999% pure vanadium target (Power V) with two levels 40 Watts, and 50 Watts; DC power applied to the 99.9999% pure carbon target (Power C) with two levels 30 Watts, and 90 Watts and deposition temperature (Temperature) with two levels 18 °C and 20 °C.

This experimental design configures eight different combinations (sample number column in Table 1) of factors and levels which replicated three times results in 24 different coatings experiments. Substrate rotation of 10 rpm, deposition time of 30 min and argon (Ar) flow rate of 30 standard cubic centimeters per minute (sccm) were kept constants as well as the samples material AISI-SAE 1045 steel. The substrates were previously mechanized and cleaned with isopropanol and acetone to remove organic compounds on the surface previously to the deposition process [3].

Table 1. 2^3 designs of experiments for coating's deposition.

Factors		Low value (-)	High value (+)
Power V (W)		40	90
Power C (W)		30	50
Temperature (°C)		18	200
Sample	**Power V (W)**	**Power C (W)**	**Temperature (°C)**
1	40	30	18
2	90	30	
3	40	50	
4	90	50	
5	40	30	200
6	90	30	
7	40	50	
8	90	50	

2.2. Roughness measurements

The roughness values of the coated samples surface were obtained under normal laboratory atmosphere conditions, temperature around 20 °C and relative humidity between 18% and 20%. For these measurements, the portable roughness tester PCE-RT 1200 was used under parameters method 9230-ST-I-27. Twelve tests were done per sample, 6 measurements in 2 different zones of the coating with the aim to assure reproducibility; also, roughness sample without coating was measured.

2.3. Electrochemical evaluation of the coatings

The electrochemical measurements made on the samples were carried out by means of an electrochemical workstation CorrTest CS350 with a sweep speed of 0.125 mV/s in a range of voltages between -300 mV to 250 mV with respect to the corrosion potential (Ecorr). Deionized water solution with 3.5% of sodium chloride (NaCl) at room temperature was used. By extrapolation of Tafel slopes, the corrosion current (nA) and the corrosion rate (mpy) from the polarization curves

3. Results and discussion

Table 2 shows the roughness average measurements of vanadium-carbon-based coatings with the 8 different deposition parameters combinations (sample column) derived from the factorial design in Table 1. Roughness of the uncoated substrates lies around 3.89 μm which corresponds to the typical value of a machined AISI-SAE 1045 steel surface' finishing [8]. Comparing this value against all the ones of the samples a reduction is exhibited except for the sample 6 which presented a roughness close to the uncoated substrate one.

From the point of view of deposition temperature, a relevant influence of it is evidenced, the samples grown at room temperature presented in average 0.8 μm less roughness than those grown at 200 °C. These results match those observed in earlier studies by Carroll, *et al.* [9] stablished that roughness of vanadium-carbon-based thin films highly depends on the deposition temperature. This reduction of roughness is attributed to the production process, the sputtered coatings are homogeneous and tends to be smooth surfaces.

The electrochemical studies done by means of potentiodynamic polarization technique and with the use of Tafel extrapolation method are summarized in Table 3. In terms of the corrosion current (Icorr), the sample 3 showed the lowest value (12.4 nA) and this is in accordance with the lowest roughness (1.333 um) as showed in Table 2. This behavior is attributed to the deposition parameters configuration in sample 3, 40 Watts of DC power applied to the vanadium target, 50 Watts of DC power applied to the carbon target and room temperature. In agreement with Pat and Korkmaz [10] the deposition temperature highly affects the coatings' properties against corrosion phenomena, the higher the deposition temperature, the higher the corrosion current density.

Table 2. Roughness measurements of VC coatings.

Sample	Results Ra (μm)	
	Measure	Standard deviation
1	2.047	0.686
2	1.593	0.427
3	1.333	0.664
4	2.609	0.718
5	2.313	0.717
6	3.649	2.150
7	1.938	0.331
8	2.912	0.490

*VC is vanadium carbide.

Table 3. Electrochemical measurements.

Sample	Icorr (nA)	Ecorr (mV)	Corrosion rate (mpy)
1	18.6	-45.40	8.497×10^{-3}
2	14.3	-13.80	6.520×10^{-3}
3	12.4	53.30	5.660×10^{-3}
4	46.3	-17.50	21.11×10^{-3}
5	54.3	-119	24.75×10^{-3}
6	880.0	-60.10	400.9×10^{-3}
7	38.2	7.05	17.40×10^{-3}
8	197.0	-120	89.76×10^{-3}

The potentiodynamic polarization curves are shown in Figure 1. The samples that were produced at room temperature tends to improve the corrosion resistance in the group of samples, this can be seen with the corrosion current values which are lower than the ones grown at 200 °C of temperature.

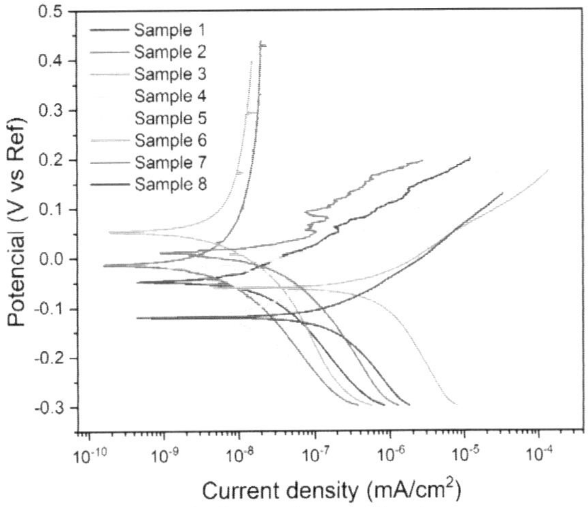

Figure 1. Potentiodynamic polarization curves.

The average scores of the corrosion current and corrosion rate at the two levels of temperature were compared to evaluate the effect of this factor. The effect of the temperature on the corrosive parameters is relevant (see Figure 2) due to the positive slope of the graph when comparing low vs high temperature, statistically the p-value is less than 0.05. Using room temperature vs using 200 °C, with the high temperature the corrosion rate increases as well as the corrosion current, the lower the deposition temperature the lower the kinetic energy of the atoms dropping the electronic flux around the sample as well as the redox reactions making the coating as noble material.

This goes in detriment of the anticorrosive response of the coating and can be attributed to the porosity and higher roughness of the samples grown at high temperature, the higher the roughness, the higher the contact area of the surface with the electrolyte and higher corrosion rates due to the possible presence of more corrosion paths towards the material, these paths promote the creation of reaction sites in which the ions exchange occurs creating the growth of pits or pitting corrosion [11-13].

Similarly, the effect of the DC power applied to the vanadium target is shown in Figure 2, as well as temperature the effect is relevant (p-value less than 0.05). The higher the power the higher the corrosion rate as well as the roughness of the sample. This tells us that increase the power from 40 Watts to 90 Watts negatively affects the desired properties of the coating. On the other hand, the effect of the power applied to the carbon target (see Figure 2) presents a reduction in corrosion rate when we applied 50 Watts and an increase in corrosion rate working with 30 Watts. This behavior is opposite to the one with the vanadium target.

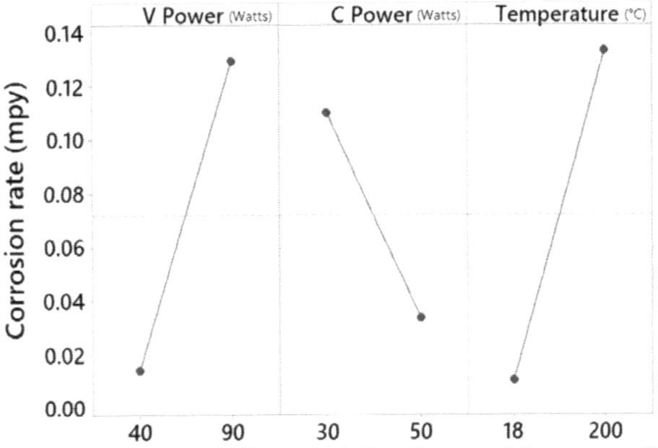

Figure 2. Main effects of the factors V power, C power, and temperature, on the corrosion rate.

These results suggest that to achieve the lowest corrosion rate as well as the lowest corrosion current, it must be considered the deposition parameters set like power applied to vanadium target at 40 Watts, power applied to carbon target at 50 Watts and room temperature. In the seek for a combination of the deposition parameters that best match the desired properties of the coatings, a superposed contours plot was made and is shown in Figure 3. To clarify this plot has two factors as actives power applied to V target and power applied to C target and temperature was fixed at room temperature (18 °C) since as we see in Figure 2 was the level of temperature that allow to have low corrosion rates and roughness; and two response variables corrosion rate and roughness.

Looking for corrosion rates low values we set a small range from 0.006 mpy to 0.01 mpy and looking for low roughness values we set a range from 1 μm to 1.5 μm, a possible window of possible values (white region on the plot) is evidenced. This result suggests that producing vanadium-carbon based coatings at room temperature to achieve good anticorrosive properties, vanadium target power must be from 42 Watts to 49 Watts and carbon target power from 46 Watts to 50 Watts.

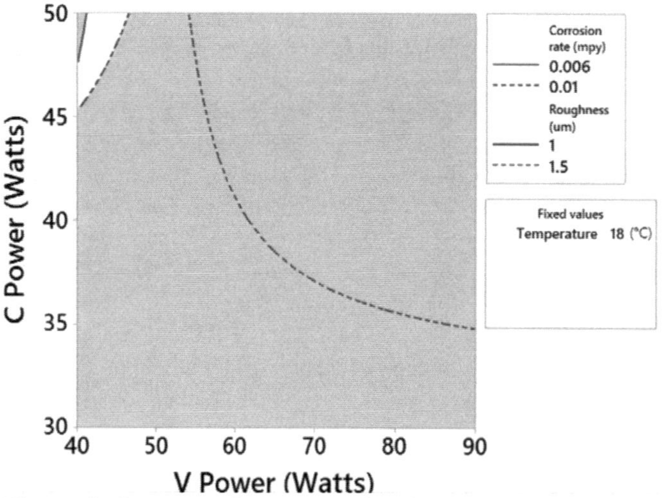

Figure 3. Superposed contours plot for power applied to C target and V target.

4. Conclusions

Corrosion analysis by means of potentiodynamic polarization curves depicts that low deposition temperature (18 °C) in general leads to better corrosion performance, for instance, in sample 2 and sample 3; also, as already discussed, these results match with lower roughness values. In contrast, when high temperature is used on substrates (*i.e.*, 200 °C) corrosion performances decreases, which suggest that vanadium carbide coatings must be deposited at room temperature and using low power applied to vanadium and carbon targets. According to superposed contours plot, to obtain optimal coatings with roughness between 1 μm to 1.5 μm and corrosion rate in the range of 0.006 mpy to 0.01 mpy, they should be deposited with similar power in both targets; being these values between 40 Watts to 48 Watts.

References

[1] Rech J, Claudin C, D'Eramo E 2009 Identification of a friction model-Application to the context of dry cutting of an AISI 1045 annealed steel with a TiN-coated carbide tool *Tribology International* **42(5)** 738

[2] Shi Z, Li X, Duan N, Yang Q 2021 Evaluation of tool wear and cutting performance considering effects of dynamic nodes movement based on FEM simulation *Chinese Journal of Aeronautics* **34(4)** 140

[3] Lontos A, Gregoriu A 2019 The effect of the deformation rate on the wall thickness of 1.5LT PET bottle during ISBM (Injection Stretch Blow Molding) process *Procedia CIRP* **81** 1307

[4] Aguzzoli C, Figueroa C A, de Souza F S, Spinelli A, Baumvol I J R 2012 Corrosion and nanomechanical properties of vanadium carbide thin film coatings of tool steel *Surface and Coatings Technology* **206(10)** 2725
[5] Leunda J, Soriano C, Sanz C, García Navas V 2011 Laser cladding of vanadium-carbide tool steels for die repair *Physics Procedia* **12** 345
[6] Orjuela G A, Rincón R, Olaya J J 2014 Corrosion resistance of niobium carbide coatings produced on AISI 1045 steel via thermo-reactive diffusion deposition *Surface and Coatings Technology* **259** 667
[7] Kelly P J, Arnell R D 2000 Magnetron sputtering: a review of recent developments and applications *Vacuum* **56(3)** 159
[8] Dhage S, Jayal A D, Sarkar P 2019 Effects of surface texture parameters of cutting tools on friction conditions at tool-chip interface during dry machining of AISI 1045 steel *Procedia Manufacturing* **33** 794
[9] Carroll J, Krchnavek R, Lunk C, Scabarozi T, Lofland S, Hettinger J 2014 An investigation of the properties of epitaxial chromium-substituted vanadium carbide thin films *Vacuum* **109** 212
[10] Pat S, Korkmaz S 2015 Nanostructured vanadium carbide thin films produced by RF magnetron sputtering *Scanning* **37(3)** 197
[11] Mariani F E, Takeya G S, Lombardi A N, Picone C A, Casteletti L C 2020 Wear and corrosion resistance of Nb-V carbide layers produced in vermicular cast iron using TRD treatments *Surface and Coatings Technology* **397** 126050
[12] Sen U 2004 Kinetics of niobium carbide coating produced on AISI 1040 steel by thermo-reactive deposition technique *Materials Chemistry and Physics* **86(1)** 189
[13] Wu X, Li G, Chen Y, Li G 2009 Microstructure and mechanical properties of vanadium carbide coatings synthesized by reactive magnetron sputtering *International Journal of Refractory Metals and Hard Materials* **27(3)** 611

Journal of Physics: Conference Series 2153 (2022) 012009 doi:10.1088/1742-6596/2153/1/012009

Structure and microstructure behavior of iron doped potassium sodium niobate powders

S Amaya-Zabala[1], A Echavarría-Isaza[2], J Tobon[3], R Roca[4], and F A Londoño-Badillo[1]

[1] Grupo de Investigación de Estado Sólido, Universidad de Antioquia, Medellín, Colombia

[2] Grupo de Investigación de catalizadores y adsorbentes, Universidad de Antioquia, Medellín, Colombia

[3] Grupo de Instrumentación Científica y Microelectrónica, Universidad de Antioquia, Medellín, Colombia

[4] Centro de Desenvolvimento de Materiais Funcionais, Universidade Federal de Sao Carlos, Sao Carlos, Brasil

E-mail: sebastian.amayaz@udea.edu.co

Abstract. In this paper, the synthesis and characterization of the potassium sodium niobate doped with iron powders have been studied. Solid-state oxide reaction sintering was used. The powders produced in this work exhibit no homogeneous microstructure, which introduced the growth of random cylindrical structures and will can contribute to the increased porosity ceramics. It was observed average particle size of 3μm, besides, also it was observed the formation of agglomerations and an increase in the size of these clusters with the increase in the amount of iron. The calcination temperature was 950 °C. This is slightly higher than other potassium sodium niobate powders systems. In addition to the physical and microstructural properties, structural properties are presented and analyzed for the first-time using Mössbauer spectroscopy as complementary technique in Fe^{3+} doped potassium sodium niobate powders. This work is important to state solid physics because establishes the influence of iron in the potassium sodium niobate system, and so the future obtaining of multifunctional materials that have piezoelectric and magnetic properties.

1. Introduction

Lead-free piezoelectric ceramics have received increasing attention from a viewpoint of environmental protection on the earth. In 2006, the European Union has already adopted the well-known directives waste electrical and electronic equipment (WEEE) and restriction of the use of certain hazardous substances in electrical and electronic equipment (RoHS), with the goal of protecting human health as well as the environment by exclusion or substitution of hazardous substances used in electrical and electronic devices[1–3] Especially potassium sodium niobate (KNN) ceramics are reported to show desirable properties for solid ultrasonic delay line application and ultrasonic transducer[4]. Therefore, tremendous efforts have been devoted to the development of KNN based lead-free ceramic to replace the commercial ceramics used for this purpose[5].

However, it is difficult obtain samples with a high density by conventional preparation and sintering [6–8]. Other problems include secondary phases appear, probably due to their high volatility of the alkali metals sodium and potassium during the sintering process and the difficulty of obtaining a fine microstructure during sintering of these materials and cause compositional fluctuations and poorer properties[6,7,9]. Thereupon, the processing and powder synthesis are critical steps in obtaining such materials, thus recently the KNN-Systems there are being obtained through “soft chemistry” methods as sol-gel, hydrothermal [10,11], and microemulsion [3]. This “soft chemistry” to prepare KNN nanoparticles should enhance the synthesis temperature, reducing the volatilization of alkali elements and densification problems compared to the traditional sintering process. Anyway, the chemical methods could not obviate the appearance of carbonates during the thermal treatment and thus high temperatures are required to synthesize adequate powders. Generally, have been obtained achievements in the process of KNN through the addition of dopants agents, such as a few types of aliovalent metals that are often used to tailor the electrical properties of KNN materials by forming point defects[10].

In the present study, KNN doped powders were prepared using conventional sintering. The main goal was to study the effect of introducing the hematite (Fe_2O_3) into the KNN system and for the first time perform Mössbauer spectroscopy measurements which have not reported on this system. Therefore, it is noteworthy in this work the novelty of investigating for the first time the magnetic behavior of a smart system such as the KNN, highly doped with Iron, through Mössbauer measures. This offers the approximate results of the placement of the iron ions in the perovskite structure, their possible oxidation state, and the appearance of new phases into de KNN structure.

2. Methodology and materials

$(K_{0,5}Na_{0,5})NbO_3 + x$ % Mass Fe, powders (KNN: xFe with x=0.5, 1.0 and 1.5) were prepared by oxide powder mixture. High-purity potassium carbonate, sodium carbonate, niobium oxide, and iron oxide (Hematite) were used as raw materials. The milling process was carried out in a Fritsch Pulverisette planetary mill using a polyvinylchloride (PVC) container, with ZrO_2 cylinders of 5 mm for 24 h in a mixture of water and isopropyl alcohol (3:1) as a medium, the resultant slurry was dried to 100 °C for 24 h and crushed in an agate mortar. Similar to process reported by Quintero et al.[12]. The powders were calcined at 950 °C for 3h, in an alumina crucible thermogravimetric (TGA), and differential thermal analysis (DTA) was carried out on KNN-Fe samples before the calcination step using TA Instruments. Model Hi-Res TGA 950 and DTA Instruments, Model 1600 DTA. Around 20 mg of powers were placed in a crucible and heated up to 950 °C with a heating rate of 5 °C /min. The measurements were performed in a flowing air atmosphere. Scanning electron microscopy (SEM) JSM-6490 was performed on the powder samples, the dust samples were scattered on a sample holder through ultrasound. The structural evolution produced by Fe addition on the perovskite structures was followed by X-ray diffraction (XRD) powder diffraction (XPert PANalytical Empyrean Series II con Detector PIXcel 3D (Co: 1.79Å) to 5 degrees per minute and room temperature.

The Mössbauer spectra were obtained at room temperature in the standard transmission geometry using a $^{57}Co/Rh$ gamma radiation source. At the time was adjusted the amount of sample for the acquisition of the Mössbauer spectrum, performed for thickness using the Mössbauer spectral analysis software (RECOIL). Since, the spectra were adjusted and analyzed with the same program by using a nonlinear least-squares fitting procedure with the Lorentzian Mössbauer lines.

3. Result and discussion

The course of the reaction between the initial precursors was analyzed by means of a thermal analysis for KNN: xFe with x = 1.0% as shown in Figure 1. In Figure 1 an exothermic peak is observed in the DTA at 315 ºC, this indicates the reaction between alkali carbonates and the Nb_2O_5 [11]. Regarding weight loss, before 200 ºC of temperature, the weight loss is 12% that can be attributed to water evaporation. The presence of water in the carbonate-oxide powders mixture is due to the hygroscopic nature of carbonates[13]. At temperature 620 ºC total mass loss of 20.2% for x = 1.0% also can be seen in the Figure 1. We supposed that the complete decomposition of carbonates was observed at 620 °C. Similar behavior was presented by another iron concentration (not reported here).

Five weight loss peaks can be observed on the TG curve, at 85 ºC- 95 °C, 160 °C, 315 °C, 480 °C, and 611 °C, associated with endothermic peaks, those high losses are attributed not only to the removal of absorbed environmental moisture water and chemical decomposition of the precursors (315 °C, 380 °C) [14]. Also, it is possible contamination with the material of the PVC bottle (160°C) used in grinding. In the same way, the endothermic peaks at 345 °C and 480 °C are consistent with the polymorphic transition of A_2CO_3 [15]. At temperatures above 930 °C, the decomposition of carbonates has been completed, therefore, the calcination temperature finish in the complete carbonate decomposition was select at ~950 °C during 3 h (the concentrations, x = 0.5% and x = 1.5% present the same behavior, which not are shown here).

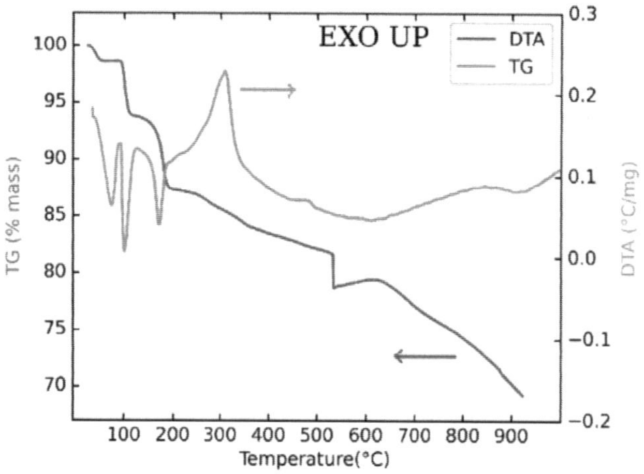

Figure 1. Thermogravimetric and differential thermal analysis (TGA/TD) of KNN doped with 1.0% Fe powder.

Figure 2 depicts the XRD pattern of KNN: x %Fe with x = 0.5% (Figure 2(a) blue spectrum), x = 1.0% (Figure 2(a) reds spectrum), and x = 1.5% (Figure 2(a) green spectrum) powders calcined at 950 °C for 3 h; graphic that represent secondary phase percent (Figure 2(b)). The powders crystallized in the orthorhombic space group Amm2 (JCPDS-ICDD 33-1270). In the same way, all peaks of the KNN: x %Fe powders are consistent with $NaNbO_3$ peaks, however, there is an increase in the number of secondary phase peaks with the increase in the percentage of iron in the KNN system (Figure 2(b) can be seen the second phase evolution in the function of the increase of iron). We suppose that above is because the Fe^{+3} quantity used in this work is greater than the solubility limit of Fe into the KNN matrix system.

In general, we can notice that with the increase of Fe^{+3} doped agent, the appearance of secondary phases in the KNN structure increase too, as can be seen in Figure 2(b), the proportional increase of secondary phases, calculated by means of Equation (1), in this we find the percentage of secondary phases is equal to the sum of the main peaks plus the sum of the secondary peaks divided by the sum of the main peaks [16] with the XRD pattern we can demonstrate that for all the concentrations, the Fe^{+3} ion are incorporated into the matrix of KNN phase. Owing to their similar sizes, the Fe^{+3} ions $(0.69\ \text{Å})$ enter Nb^{+5} $(0.68\ \text{Å})$ sites and act as an acceptor dopant. Oxygen vacancies will be introduced into the lattice to compensate for the charge imbalance with the doping of acceptor Fe^{+3}.

$$\%\ \text{seconday Phases} = \frac{\sum \text{1st peaks} + \sum \text{2nd peaks}}{\sum \text{1st peaks}}. \quad (1)$$

Figure 3 shows the SEM micrographs for each of the concentrations studied of KNN-Fe powders calcined at 950 °C. As can be seen from Figure 3, the SEM micrographs (with x = 1.0% and x = 1.5%) present an irregular contour that according to some authors indicating the reaction is not completed [17,18]. This observation can indicate that x = 0.5%, the Fe concentration is around the solubility limit. Also, this result confirmed the appearance of the secondary phases with the iron increase, however, the increase of Fe^{+3} concentration should increase the calcination temperature, nevertheless, the TDA and TG analysis does not show it.

It is evident changes in grain size and morphology with the increase in the amount of x% Fe_2O_3 in the $(K_{0,5}Na_{0,5})NbO_3$ system. The grain size decreases, and agglomerations are perceived whose size ranges from $\approx$ 3 µm to 40 µm. Alike, the change in morphology can be attributed to the high concentration of iron, which can be confirmed by the formation of secondary phases observed in the DRX, see Figure 2. This occurs mainly because of the non-assimilation of iron in the structure, and it is persistence of Hematite in the ceramic system. We propose that this hematite excess is carried to the grain contours, leading to a change in the electrical properties of the KNN system [18]; which could lead to an increase in magnetic properties as other studies showed [19]. This observation plays an important role in the experimental physical because it allows thinking to fabricate Fe doped KNN samples and your application in multifunctional materials.

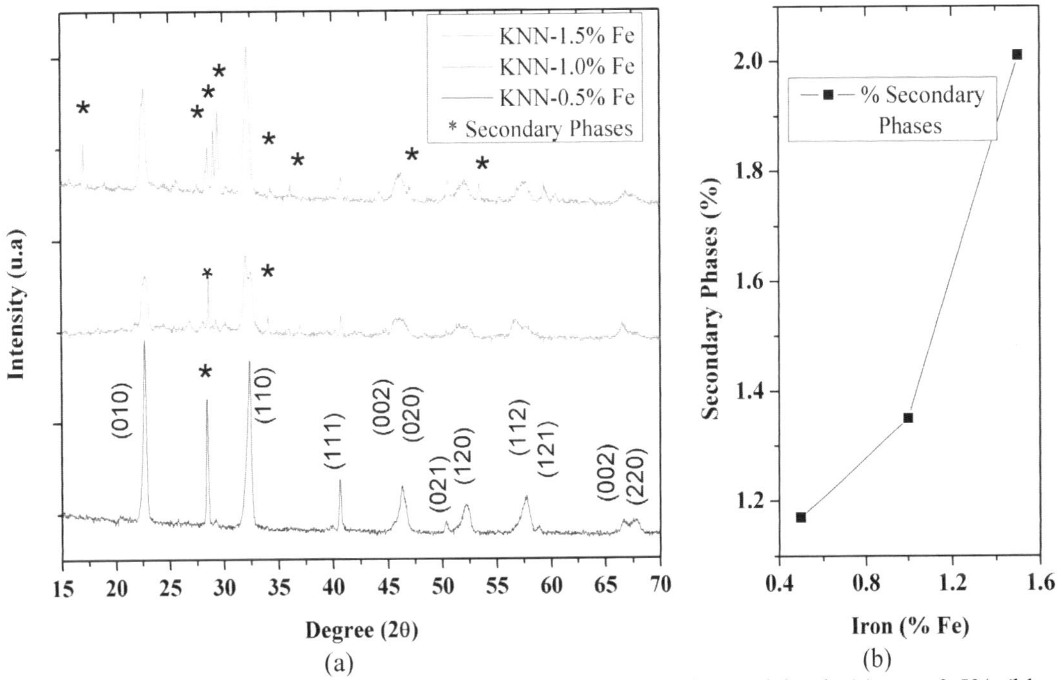

Figure 2. Pattern of X-ray diffraction of KNN: x %Fe powders calcined. (a) x = 0.5% (blue spectrum), x = 1.0% (red spectrum), x = 1.5% (green spectrum) and (b) Secondary phases percent of KNN: x %Fe.

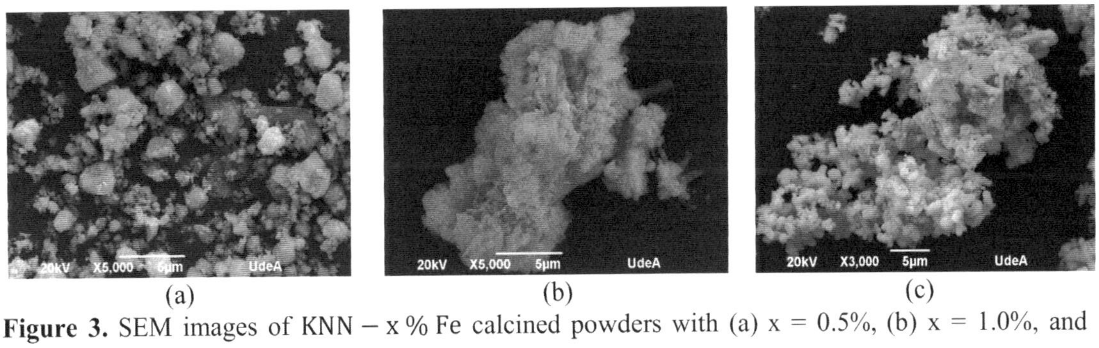

Figure 3. SEM images of KNN – x % Fe calcined powders with (a) x = 0.5%, (b) x = 1.0%, and (c) x = 1.5%.

Unpublished Mössbauer spectroscopy measurements of the system KNN:xFe can be seen in Figure 4; x = 0.5% (Figure 4(a)), x = 1.0% (Figure 4(b)), and x = 1.5% (Figure 4(c)) powders calcined at 950 °C. The most likely oxidation states available in the Mössbauer spectrum are for Fe^{+3} where the structural location could be at octahedral (A) or tetrahedral (B) sites, which indicates in solid state physics a high assimilation of iron in the Perovskite structure (ABX_3). On the other hand, Fe^{+3} increase produces an increase in secondary phases (Figure 2 (b)), however, the oxidation and coordination states are preserved [18,20] (Table 1). This occurs mainly because of the non-assimilation of iron in the structure, and it is persistence of Hematite in the ceramic system. We propose that this hematite excess is carried to the grain contours, possibly decreasing the electrical properties of the KNN system [21].

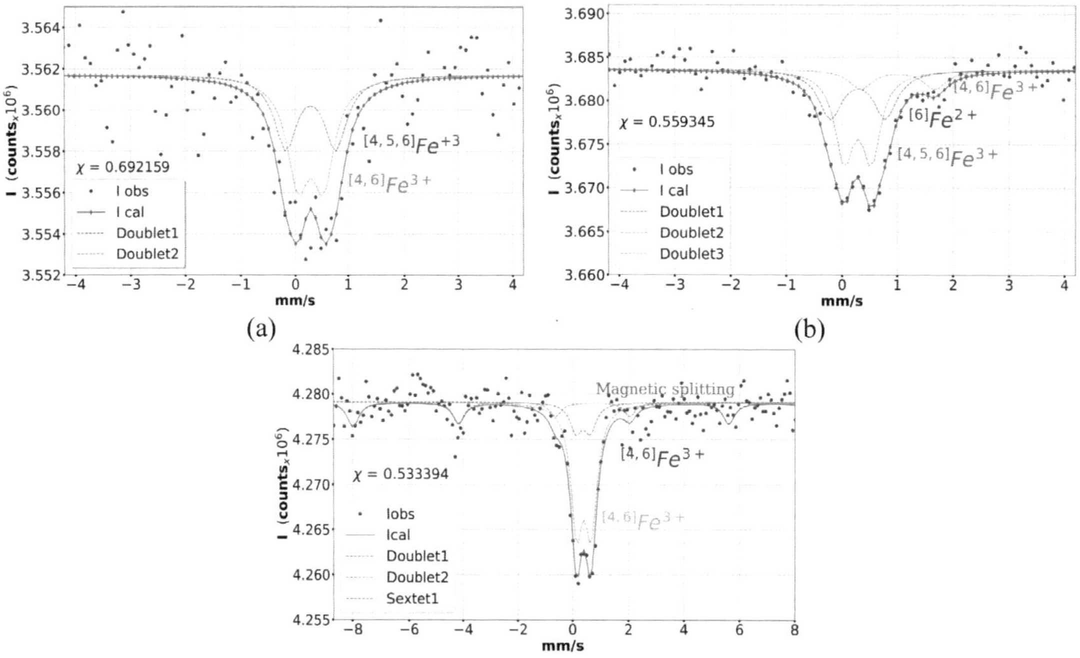

(c)

Figure 4. Mössbauer absorption spectrum of KNN – x % Fe powders. (a) x = 0.5%, (b) x = 1.0%, and (c) x = 1.5%.

Table 1. Abstract table of Mössbauer absorption spectrum of KNN – x % Fe powders.

Mössbauer data KNN - x% Fe						
x	Sites	χ^2	Cs (mm/s)	D (mm/s)	Coordination	Populations (%)
0.5	doublet1	0.483373	0.318	0.940	Fe^{+3} [4,5,6]	56
	doublet2		0.308	0.510	Fe^{+3} [4,6]	44
1.0	doublet1	0.559345	0.300	0.950	Fe^{+2} [4,6]	39
	doublet2		1.000	1.310	Fe^{+2} [6]	14
	doublet3		0.297	0.490	Fe^{+3} [4,6]	47
1.5	doublet1	0.533394	0.387	0.499	Fe^{+3} [4,6]	13
	doublet2		0.384	0.500	Fe^{+3} [4,6]	57
	sextet1		0.592			29

4. Conclusions

The processing of KNN doped Fe needs extreme control of many parameters starting since milling process to rates of heating. The calcination temperature is given at 950 °C, where the carbonates have been completely decomposed. The appearance of secondary phases in the system are influenced by iron concentration, showing right agreement phase with orthogonal structure KNb_2O_3 (PDF32- 0822) and the Mössbauer spectrum, evidence the hematite was incorporated completely into the system, manifesting an occupation at the octahedral and tetrahedral sites within the perovskite, and for the maximum iron concentration, there is a magnetic manifestation in a sextet profile.

The results obtained in this work allow in solid-state physics to begin the densification process of KNN:xFe for analysis of piezoelectric and magnetic properties in order to find applications as multifunctional materials.

It was found that the incorporation of Fe in KNN increases the calcination temperature in powders obtained by a mixture of oxides. In state-solid physics, those behavior is important because could increase stoichiometric alterations in synthesis and densification process due to that potassium and sodium are moderately volatile elements.

Acknowledgments
This research was supported by Colombian Agencies: CODI-Universidad de Antioquia (Estrategia de Sostenibilidad 2020-2021 de la Universidad de Antioquia and Project: We are very thankful to the Grupo de Investigación de Catalizadores y Adsorbentes for allowing us the use of their XRD and DTA/TG instruments.

References

[1] Li, Jing-Feng and Wang, Ke and Zhu, Fang-Yuan and Cheng, Li-Qian, Yao F-Z 2013 (K, Na)NbO 3 -based lead-free piezoceramics: fundamental aspects, processing technologies, and remaining challenges *J. Am. Ceram. Soc.* **96** 3677
[2] Tsydenova O, Bengtsson M 2011 Chemical hazards associated with treatment of waste electrical and electronic equipment *Waste Manag.* **31** 45
[3] Kosec M, Malič B, Benčan A, Rojac T 2008 KNN-based piezoelectric ceramics *Piezoelectric and Acoustic Materials for Transducer Applications* ed Safari A, Akdoğan E K (Boston: Springer) p 81
[4] Oglezneva S A, Kachenyuk M N, Kulmeteva V B, Ogleznev N B 2017 Structure and characteristics of functional powder composite materials obtained by spark plasma sintering *IOP Conf. Ser. Mater. Sci. Eng.* **218(1)** 012014:1
[5] Chen B, Liang P, Wu D, Zhao X, Qiao X, Peng Z, Wei L, Chao X, Yang Z 2019 High-efficiency synthesis of high-performance K0.5Na0.5NbO3 ceramics *Powder Technol.* **346** 248
[6] Hollenstein E, Damjanovic D, Setter N 2007 Temperature stability of the piezoelectric properties of Li-modified KNN ceramics *J. Eur. Ceram. Soc.* **27** 4093–7
[7] Fu Z, Yang J, Lu P, Zhang L, Yao H, Xu F, Li Y 2017 Influence of secondary phase on polymorphic phase transition in Li-doped KNN lead-free ceramics *Ceram. Int.* **43** 12893
[8] Ryu J, Choi J J, Hahn B D, Park D S, Yoon W H, Kim K Y 2007 Sintering and piezoelectric properties of KNN ceramics doped with KZT *IEEE Trans. Ultrason. Ferroelectr. Freq. Control* **54** 2510
[9] Pan Z, Chen J, Fan L, Zhang J, Zhang S, Huang Y, Liu L, Fang L, Xing X 2015 Enhanced piezoelectric properties and thermal stability in the (K0.5Na0.5)NbO3:ZnO lead-free piezoelectric composites *J. Am. Ceram. Soc.* **98** 3935
[10] Wu J, Xiao D, Zhu J 2015 Potassium-sodium niobate lead-free piezoelectric materials: past, present, and future of phase boundaries *Chem. Rev.* **115** 2559
[11] Malič B, Kupec A, Kosec M 2013 Thermal analysis *Chemical Solution Deposition of Functional Oxide Thin Films* ed Schneller T, Waser R, Kosec M, Payne D (Vienna: Springer) p 163
[12] Quintero M C, Rincón M, Osorio-Guillén J M, López D, Londoño F A 2019 Grinding methods effects on the synthesis of Potassium-Sodium Niobate powders by oxide mixing *Tecnológicas* **22(46)** 15
[13] Kuu W Y, Chilamkurti R, Chen C 1998 Effect of relative humidity and temperature on moisture sorption and stability of sodium bicarbonate powder *Int. J. Pharm.* **166** 167
[14] Vojinovi V, Cabral J M S, Fonseca L P 2006 Real-time bioprocess monitoring: part I: in situ sensors *Sensors and Actuators B: Chemical* **114(2)** 1083
[15] Rubio-marcos F, Romero J J, Martín-gonzalez M S, Fernández J F 2010 Effect of stoichiometry and milling processes in the synthesis and the piezoelectric properties of modified KNN nanoparticles by solid state reaction *Journal of the European Ceramic Society* **30(13)** 2763
[16] Karlak R F, Burnett D S 2002 Quantitative phase analysis by x-ray diffraction *Anal. Chem.* **38** 1741
[17] Chen K, Zhang F, Li D, Tang J, Jiao Y, An L 2016 Acceptor doping effects in (K0.5Na0.5)NbO3 lead-free piezoelectric ceramics *Ceram. Int.* **42** 2899

[18] Xu F, Chen J, Lu Y, Zhang Q, Zhang Q, Zhou T, He Y 2018 Exploration on the origin of enhanced piezoelectric properties in transition-metal ion doped KNN based lead-free ceramics *Ceram. Int.* **44** 16745
[19] Zhou M, Lu X, Liu L, Wang J, Lu D, Huang F, Zhu J, Cheng P, Wang Q 2020 Room temperature multiferroic properties and polymorphic phase transition-induced noticeable magnetodielectric anomalies in Fe/Co co-doped (K0.52Na0.48)NbO3 ceramics *J. Alloys Compd.* **836** 155519
[20] Wang Y, Zhu W, Sun Q, Tan L 2019 Effects of A/B-site dopants on microstructure and domain configuration of potassium sodium niobate lead-free piezoelectric ceramics *J. Alloys Compd.* **787** 407
[21] Murad E, Cashion J 2004 *Mössbauer Spectroscopy of Environmental Materials and Their Industrial Utilization* (New York: Springer)

Development of an integrated cooling system for motorcycle helmet by physical study of temperature variations and heat transfer

S F Zambrano-Becerra[1], P M Galvis-Sanchez[1], N Y Perez-Rangel[1], E Florez-Solano[1], and E Espinel-Blanco[2]
[1] Grupo de Investigación en Ingenierías Aplicadas para la Innovación, Gestión y Desarrollo, Universidad Francisco de Paula Santander, Seccional Ocaña, Colombia
[2] Grupo de Investigación en Tecnología y Desarrollo en Ingeniería, Universidad Francisco de Paula Santander, Seccional Ocaña, Colombia

E-mail: nperezr@ufpso.edu.co, pmgalviss@ufpso.edu.co

Abstract. In Colombia, the most widely used means of transport today are motorcycles, which have become increasingly numerous, bearing in mind that they are subject to laws and regulations imposed by the country's mobility, transit and transport agencies, the use of helmets is mandatory for drivers and passengers, safety measures are monitored, the hull must be certified and meet the required technical standards; whereas its role is to protect people in the event of accidents, regulations require that the helmet be completely closed to protect the entire head and chin; the design of the helmet allows air entry and there is no concentration of temperature inside, all this is done by implementing air inlet and outlet ducts, which circulate air when the motorcycle is in motion, unfortunately this does not happen due to the accumulation of temperature in the back of the helmet that makes the user feel tired and uncomfortable. This research proposes the development of a prototype portable cooling system for motorcycle helmets by the physical principle of heat transfer, by using Peltier cells, to have low production cost, optimal operation, and low energy consumption thanks to natural air flow.

1. Introduction

In Colombia, statistics from "Asociación Nacional de Empresarios de Colombia (ANDI)" and "Federación Nacional de Comerciantes (FENALCO)" confirm that sales of motorcycles at the national level in Colombia are increasing, in such a way that compared to the years 2020 and 2021 an increase of 44% [1], this indicates that the number of motorcycles circulating is much higher; therefore, in order to become a creditor of any motorcycle-type vehicle, the rules established by the "Ministerio de Transporte", Colombia, for circulation within a given territory must be complied with [2].

Helmets must comply with the technical regulations specified in "Norma Técnica Colombiana (NTC)" NTC 4533 [3], which dictates the minimum parameters that the helmet must have to be fit in circulation and not be fined for infringing the established law [4]. Within these specifications are that they should be closed helmets that protect the entire skull and chin [5,6]. On the other hand, the "Instituto de Hidrología, Meteorología y Estudios Ambientales de Colombia (IDEAM)" [7], show in their statistics how a substantial increase in temperature is seen at the national level, especially in tropical and warm zones, this triggers the motorcycle population to be more exposed to solar radiation, which means interacting with excessive heat and sweating, causing a feeling of despair in the upper part of the head due to the accumulation of heat in this area of the helmet, caused by the deficiency of the ventilation systems they bring, causing the temperature to rise in various areas and achieving a bad experience in the motorcycle driver [8,9].

To build the air conditioning device for motorcycle helmet, you have to take into account many of the principles that govern the exchange of heat, several authors have put forward various theories based on Peltier effects devices and specifically on heat pumps, so that the first helmet in the world with air conditioning works with one of these pumps, based on such simple concepts as heat; which is a form of energy exchange between the system and its surroundings due to a temperature difference, heat-free thermodynamic systems, possess energy, and one of the ways in which they exchange this energy is heat.

Heat transfer, which is a process through which energy is exchanged in the form of heat between different bodies, or between different parts of the same body that are at different temperatures. Peltier effect; its functionality is produced when a current is passed through a circuit composed of different materials whose joints are at the same temperature, an inverse effect is produced, in this case, heat is absorbed in one joint and the temperature is released in the other and finally, which is a physical magnitude that determines or reflects the amount of heat of an object, of the environment or of a body [10]. This research proposes the development of a prototype test of a portable cooling system through the physical study of temperature variations for laboratory tests at scale to provide the basis for the construction of a final prototype reproducible on a larger scale and giving the user a sense of comfort when moving in this mode of transport, where the perception of heat inside the helmet decreases, without neglecting the low cost, optimal performance and low energy consumption with the use of Peltier cells and the use of natural air flow when driving the motorcycle [11].

2. Methodology and materials

For the construction of the prototype of the integrated cooling system for the motorcycle helmet, the following methodology shown in Figure 1 is proposed, which contemplates the phases of the process to be followed to achieve the realization of the prototype.

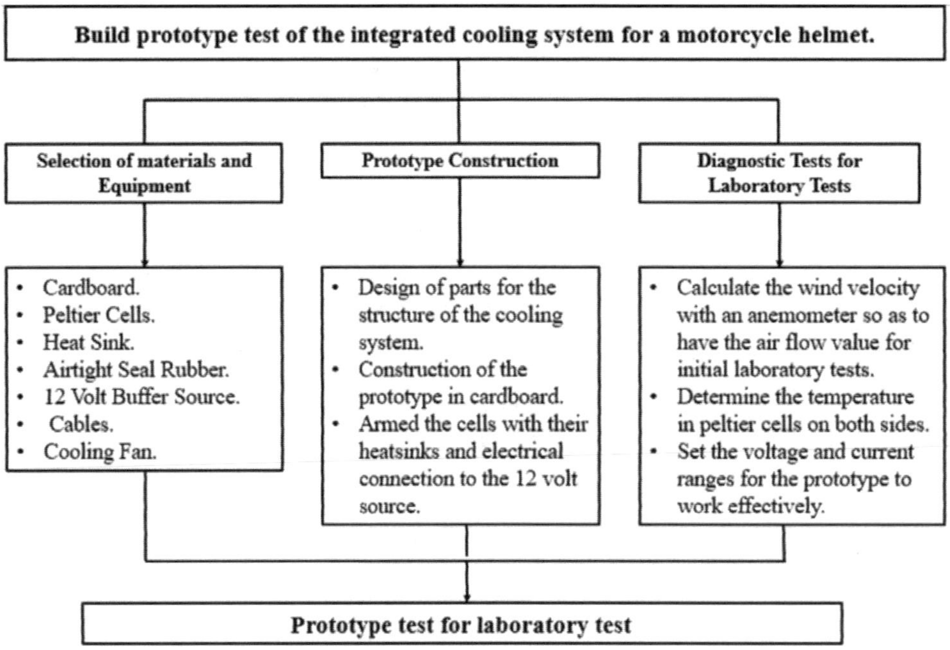

Figure 1. Methodology for the construction of the prototype air-conditioning test for motorcycle helmet.

This research will be carried out in 3 stages as detailed in Figure 1; the first thing that will be carried out is the selection of the materials and equipment necessary to be able to carry out the test prototype, among which we have: cardboard for the base and structure of the system, Peltier cells for cooling the air entering the hull, heat sink allowing heat transfer from the hot part of the cell, airtight seal rubber to prevent leakage, 12 volt regulating source for power supply to the system, cables for electrical connections and cooling fan.

The second stage is the prototype construction that will be carried out from the previous design in the SolidWorks software [12], where each of the components that will be part of the system structure will be drawn in cardboard, Next to this will be installed Peltier cells and heat sinks to the 12 volt source and all electrical connections; Finally, diagnostic tests for laboratory tests, consisting of the calculation of the wind speed to have the air flow value for initial laboratory tests, shall be carried out in stage three; as it is to determine the temperature in the Peltier cells on both sides, and to set the voltage and current ranges for the prototype to function effectively.

3. Results

Figure 2 shows the initial prototype designed and assembled in SolidWorks, Figure 2(a) show front view, Figure 2(b) show side view, including its location on the rider's helmet, considering that a portable system is proposed, ergonomic that generates a state of comfort in the pilot when driving and economical with the use of Peltier cells [13].

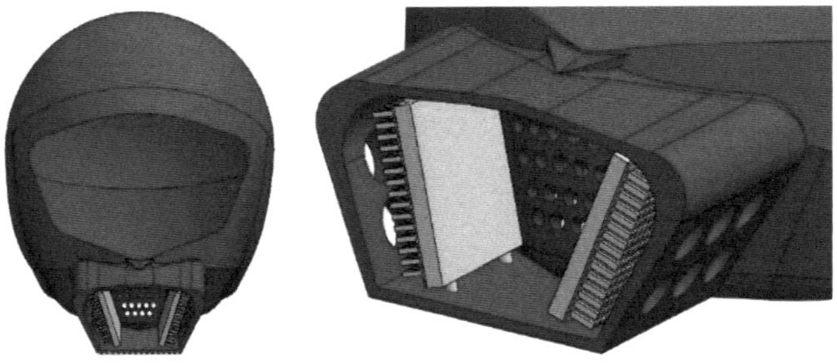

(a) (b)

Figure 2. Integrated cooling system prototype design for motorcycle helmet; (a) front view; (b) side view [12].

The whole process was based on the application of physical study that includes the behavior of air flow temperatures as it enters the hull, in the same way the heat transfer to enable Peltier cells to carry out what was originally intended, which is to be able to reduce the air temperature at the entrance of the hull, factors such as the generation of heat on the opposite side of the cell were also taken into account so as not to affect the air-conditioning system, However, there are sufficient grounds to be able to start carrying out laboratory tests leading to the final construction of the prototype taking into account all these variables mentioned above.

After adjustments to make it a system that can be used in the largest number of helmets on the market, changes were made to the prototype design to fulfill this function, the new design in the SolidWorks software can be seen in Figure 3, Figure 3(a) show front view, Figure 3(b) show side view, which will be the model for the construction of the test prototype and make small-scale laboratory tests before configuring final details as hot temperature variables produced by Peltier cells. The prototype test already built can be seen in Figure 4, Figure 4(a) show front view, Figure 4(b) show side view, was made of cardboard and was conditioned Peltier cells with their respective heat sinks and rubber to prevent leaks in the integrated air conditioning system.

Figure 5 shows an air flow simulation in the SolidWorks software that tells us that the output distribution for the environment such as the air entering the hull is uniform, Figure 5(a) show simulation start, Figure 5(b) show final simulation, which is convenient and points towards the efficiency of the prototype.

Journal of Physics: Conference Series **2153** (2022) 012010 doi:10.1088/1742-6596/2153/1/012010

In Figure 6 the electrical connections to the 12-volt regulating source for the correct operation of the Peltier cells and heat sinks can be observed, finally, Figure 7 shows how the air flow test with a thermo-anemometer is performed to define important parameters such as the actual air flow simulation for subsequent laboratory tests, of which there are flow values according to the speed travelled on the motorcycle, these values can be seen in Table 1.

The actual air flow velocity values needed to simulate in the laboratory were obtained, this will allow the project to be completed to the extent that the prototype can be realized on a larger scale in which tests are carried out on motorcyclists and in this way to evaluate factors such as the satisfaction and the impact generated in the community to which this project is directed, taking into account their innovation, low price and ergonomics.

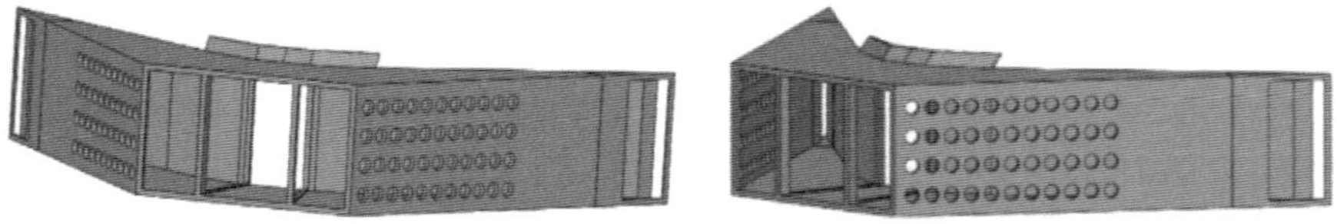

(a) (b)

Figure 3. Final design of the prototype integrated motorcycle helmet cooling system; (a) front view; (b) side view.

(a) (b)

Figure 4. Laboratory test prototype; (a) front view; (b) side view.

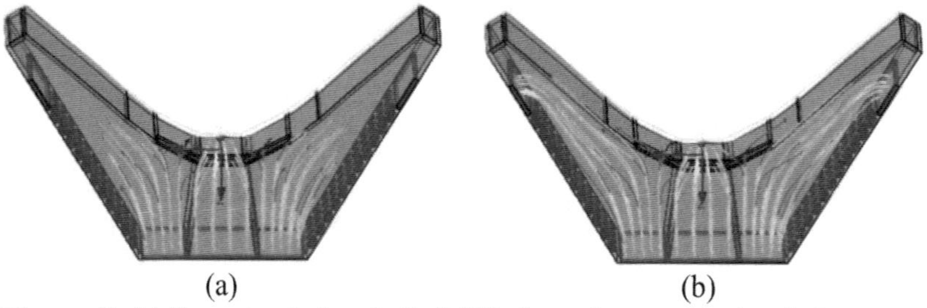

(a) (b)

Figure 5. Airflow simulation in SolidWorks software; (a) simulation start; (b) final simulation.

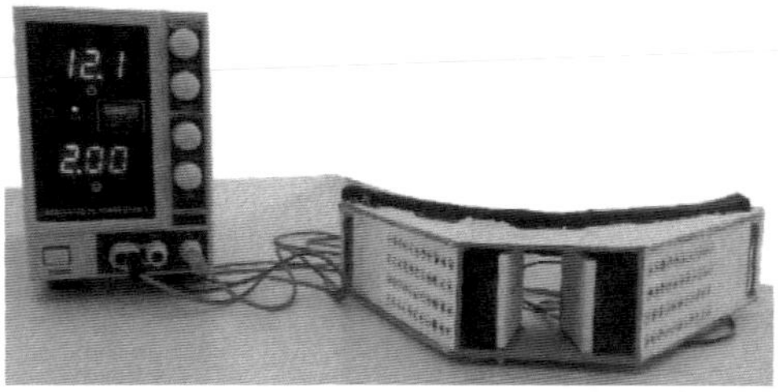

Figure 6. Electrical connections to the 12-volt control source prototype air conditioning test for motorcycle helmet.

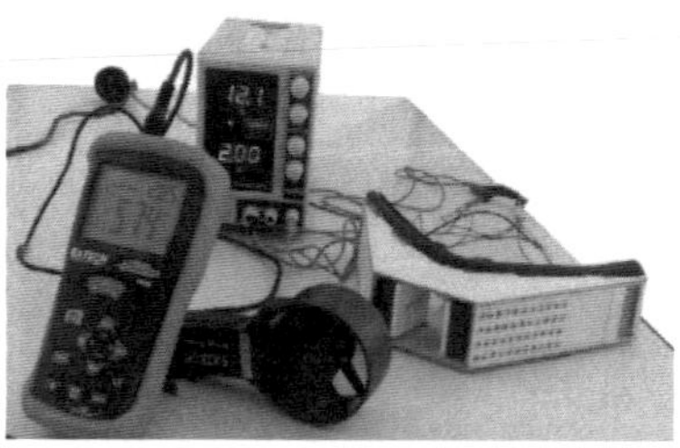

Figure 7. Air flow test with a thermo-anemometer.

Table 1. Velocities collected with the thermo anemometer.

Speed of the motorcycle (Km/h)	Speed anemometer (m/s)
20	5.55
40	11.11
60	16.66
80	22.22

4. Conclusions

The prototype test of an integrated motorcycle helmet cooling system was produced, applying the physical principles corresponding to heat transfer and temperature variations in the air flows entering the helmet, allowing this device to feel thermal comfort when driving; this initiates the conduct of small-scale laboratory tests to establish and configure the system variables to be able to build the robust prototype that will be tested on a larger scale on motorcyclists where the efficiency and effectiveness of the system is verified integrated climate control for motorcycle helmet.

The manufactured system is portable and can be upgraded to 95% of the closed helmets that comply with the legal regulations in force at national level; it is also ergonomic and economical which makes it accessible to any stratum of the motorcycle population because of its low price, this eradicates in the implementation of Peltier cells as cooling technology in the air conditioning system.

Wind speed values were recorded to simulate in the laboratory, which will make it possible to carry out the initial small-scale tests and thus open the way to future modifications needed to achieve the final prototype that can be acquired by motorcyclists.

References

[1] Budiprasojo A, Entikaria-Rachmanita R 2019 Active passive thermoelectric system helmet for personal comfort *Proceedings of The Second International on Food and Agriculture* **2** 79

[2] Merienne R, Lynn J, McSweeney E, O'Shaughnessy S M 2019 Thermal cycling of thermoelectric generators: the effect of heating rate *Applied Energy* **237** 671

[3] Instituto Colombiano de Normas Técnicas y Certificación (ICONTEC) 2017 *Cascos Protectores y sus Visores para Conductores y Acompañantes de Motocicletas, Motociclos y Motocarros, Norma Técnica Colombiana, NTC 4533* (Colombia: Instituto Colombiano de Normas Técnicas y Certificación)

[4] Das P, Tiwari P 2017 Thermal degradation kinetics of plastics and model selection *Thermochimica Acta* **654** 191

[5] Gutiérrez J A, Camas-Anzueto J L, Gómez-Pérez J, Pérez-Patricio M, Ríos-Rojas C 2019 Comparación del gradiente de temperatura en espacios cúbicos controlados con celda Peltier en función a dos aislantes térmicos *Pistas Educativas* **41(134)** 1

[6] Prakash A, Kajava A R, Jose J K, Shalimar M, Girish A R 2020 Design of temperature controlled riding gear *International Journal of Research in Engineering, Science and Management* **3(8)** 164

[7] Ghani S, ElBialy E M A A, Bakochristou F, Gamaledin S M A, Rashwan M M 2017 The effect of forced convection and PCM on helmets' thermal performance in hot and arid environments *Applied Thermal Engineering* **111** 624

[8] Remeli M F, Bakaruddin N E, Shawal S, Husin H, Othman M F, Singh B 2020 Experimental study of a mini cooler by using Peltier thermoelectric cell *Journal of Physics: Conference Series* **788(1)** 012076:1

[9] Mishra A, Kumar M 2018 Peltier thermoelectric cooling module *Global Journal for Research Analysis* **7(2)** 6

[10] Whitt R, Nafis B, Huitink D, Yuan Z, Deshpande A, Narayanasamy B, Luo F 2019 Heat transfer and pressure drop performance of additively manufactured polymer heat spreaders for low-weight directed cooling integration in power electronics *18th IEEE Intersociety Conference on Thermal and Thermomechanical Phenomena in Electronic Systems* (United States of America: Institute of Electrical and Electronics Engineers)

[11] Dizaji H S, Jafarmadar S, Khalilarya S, Pourhedayat S 2019 A comprehensive exergy analysis of a prototype Peltier air-cooler; experimental investigation *Renewable Energy* **131** 308

[12] Dassault Systèmes 2020 *SOLIDWORKS 2020* (United States of America: Dassault Systèmes)

[13] Zambrano-Becerra S F, Galvis-Sánchez P M, Perez-Rangel N Y 2020 Design of an integrated cooling system by means of thermal variation flows *Journal of Physics: Conference Series* **1708(1)** 012018:1

Physical implementation of optimal control over a direct current

D F Devia[1], **F Mesa**[1], **and D M Devia**[1]

[1] Departamento de Matemáticas, Universidad Tecnológica de Pereira, Pereira, Colombia

E-mail: dfdevia@utp.edu.co, femesa@utp.edu.co

Abstract. In the next manuscript we will show an interesting physical application of the technique of optimal control over a direct current, more exactly on a direct current motor. For this, the mathematical support or scaffolding that allows to establish the model of the problem will be presented in the development of the document, which contains a basic structure supported in a general dynamic system (linear or non-linear) and an objective function that will depend on the system. Although in essence the optimal control problems correspond to infinite spaces, the finite-dimensional case will be analyzed initially and then made an extension to the case of a quadratic linear regulator for an interval of infinite duration. The results of the implementation of this optimal control technique on a direct current motor will be analyzed by means of Bole diagrams that will represent the frequency response of the system and a follow-up diagram of set points and rejection of disturbances.

1. Introduction

Optimal control is associated with optimization problems that involve dynamic systems and seek to find a control vector that minimizes an objective function, whose value will be determined by the behavior of the system (in a sense we say that it is a measure of the quality of said behavior). Suppose then that we have a nonlinear dynamical system described by a vector $x(t)$ of n-dimensional states with a known initial state $x(0)$, an m-dimensional control vector $u(t)$ and a vector function $f(z)$ doubly differentiable analytically continuous.

The control problem consists of choosing the control trajectories $u^*(t)$, in a closed interval $[0, T]$, where T is the final time that can be set, which will help to minimize a cost function J, which depends on two real and scalar functions $L(x(t), u(t), t)$ and $\Phi(x(T), T))$ in general nonlinear. Specifically, the cost function J must be defined as the sum of the function $\Phi(x(T), T)$ and a definite integral over the optimization interval $[0, T]$ whose integrand is the function $L(x(t), u(t), t)$. This will allow us to ensure that at the final moment T the state $x(t)$ will approach some objective state, and thus guarantee that excessive efforts are not made in the control action without the state deviated from the path [1].

This technique (that of optimal control) is introduced in highly complex problems involving infinite-dimensional spaces, since basically an extension of the properties of finite-dimensional spaces is carried out to infinite-dimensional spaces; such is the case of directional derivatives and differentiable manifolds [2].

The objective of this manuscript is to make a physical implementation of the optimal control that is of great interest to the scientific community which includes a direct current motor (DC motor) with induced current, which are of great importance since almost all power electronics systems include them in their assemblies.

2. Mathematical modeling

A general nonlinear dynamic system is described by its equations in the state space Equation (1) [3].

$$\dot{x} = f(x(t), u(t), t), \tag{1}$$

where $x(t), u(t)$ and $f(z)$ are as described in the previous section; that is to say, $x(t)$, is the vector of n-dimensional states with a known initial state $x(0)$, an m-dimensional control vector $u(t)$and a vector function $f(z)$ doubly differentiable analytically continuous. The objective of the optimal control will be to find the control trajectories $u^*(t)$, with $0 \leq t \leq T$, where T is the final time and which can be fixed, which minimizes the cost function, given by Equation (2) [3].

$$J = \Phi(x(T), T) + \int_0^T L(x(t), u(t), t)dt. \tag{2}$$

Both the function $L(x(t), u(t), t)$ and the function $\Phi(x(T), T)$ are real and scalar functions, generally non-linear. The function $\Phi(x(T), T)$ allows us to ensure that at the final instant T the state $x(t)$ will approach some objective state, while the integral ensures that excessive efforts will not be made on the optimization interval in the control action, but without the state deviating from the trajectory. Now Equation (2) is considered with an infinite duration interval for this, the quadratic linear regulator is considered first in a finite interval. be the system represented as equation of states. Equation (3) [3].

$$\dot{x} = A(t)x(t) + B(t)u(t). \tag{3}$$

The cost to minimize function is a quadratic function of the state and control variables, Equation (4) [3].

$$J = \frac{1}{2}x^T(T)S(T)x(T) + \frac{1}{2}\int_0^T [x^T(t)Q(t)x(t) + u^T(t)R(t)u(t)]dt, \tag{4}$$

where it is assumed that $S(T), Q(T)$ and $R(T)$ are real, symmetric, and positively defined matrices. We will assume that the states and control are not bounded and that the final state $x(T)$ is free, with T being the supposedly known end time. Next, an analogy is established for the quadratic case Equation (4) with the general case Equation (2).

In this sense the function that penalizes the final state Φ $(x\ (T), T)$ is now described by the quadratic term $\frac{1}{2}$ $x^T(T)$ $S(T)$ $x(T)$, while the function that penalizes the states and control actions in the intermediate states $L(x\ (t), u\ (t), t)$ is now simply $\frac{1}{2}$ $[x^T(t)Q(t)x(t) + u^T(t)R(t)u\ (t)]$. To solve the optimal control problem, Lagrange multipliers will be used, and the Hamiltonian will be defined as follows, Equation (5) [3,4].

$$H\big(x(t), u(t), \lambda(t)\big) = \frac{1}{2}[x^T(t)Q(t)x(t) + u^T(t)R(t)u(t)] + \lambda^T(A(t)x(t) + B(t)u(t)). \tag{5}$$

Therefore, Equation (4) is written as Equation (6) [3].

$$\bar{J} = \frac{1}{2}x^T(T)S(T)x(T) + \frac{1}{2}\int_0^T [H\big(x(t), u(t), \lambda(t)\big) - \lambda^T(t)x(\dot{t})]dt. \tag{6}$$

The necessary conditions are Equation (7) to Equation (9).

$$\dot{\lambda}^*(t) = -\frac{\partial H}{\partial x(t)} = -Q(t)x^*(t) - A^T\lambda^*(t), \tag{7}$$

$$\dot{x}^*(t) = \frac{\partial H}{\partial \lambda(t)} = A(t)x^*(t) + B(t)u^*(t), \tag{8}$$

$$0 = \frac{\partial H}{\partial u(t)} = R(t)u^*(t) + B^T(t)\lambda^*(t). \tag{9}$$

Whose boundary conditions are Equation (10) [3].

$$x(0) \rightarrow \text{Known}, \tag{10}$$

To solve the system of differential equations represented by Equation (7) to Equation (9) first from the control action of the stationary Equation (8), we solve for $u^*(t)$, therefore, the control law will be given by Equation (11) [3]:

$$u^*(t) = -R^{-1}(t)B^T(t)\lambda^*(t). \tag{11}$$

Note that R^{-1} can always be calculated since by definition R(t) is positively defined [4]. Given the final condition Equation (10) for the Lagrange multiplier [5], it is assumed as a possible solution that an S (t) function exists throughout the interval Equation (12) and Equation (13) [3].

$$\lambda^*(t) = S(t)x^*(t), \tag{12}$$

$$\dot{\lambda}^*(t) = \dot{S}(t)x^*(t) + S(t)\dot{x}^*(t). \tag{13}$$

Equating the Equation (13) and Equation (7), we obtain Equation (14) [3].

$$\dot{\lambda}^*(t) = -Q(t)x^*(t) - A^T\lambda^*(t) = \dot{S}(t)x^*(t) + S(t)\dot{x}^*(t). \tag{14}$$

Now, Equation (8), Equation (11) and Equation (12) are replaced in Equation (14) [3], we obtain Equation (15).

$$-Q(t)x^*(t) - A^T(t)S(t)x^*(t) = \dot{S}(t)x^*(t) + S(t)[A(t)x^*(t) + B(t)u^*(t)] = \dot{S}(t)x^*(t) + S(t)[A(t)x^*(t) + B(t)R^{-1}(t)B^T(t)S(t)x^*(t)]. \tag{15}$$

Since the previous equation must be met for an arbitrary value of $x^*(t)$, it can be eliminated from the previous Equation (16) [3].

$$\dot{S}(t) + S(t)A(t) + A^T(t)S(t) - S(t)B(t)R^{-1}(t)B^T(t)S(t) + Q(t) = 0. \tag{16}$$

Equation (16) is known as Riccati's dynamic continuous equation with boundary condition S(t) [6]. Once S(t) is known, the closed loop control law can be applied by substituting Equation (12) into Equation (11), we obtain Equation (17) [3].

$$u^*(t) = -R^{-1}(t)B^T(t)S(t)x^*(t) = -K(t)x^*(t). \tag{17}$$

Being K(t) the state feedback matrix, variable over time, Riccati S(t) solution depends only on the matrices of the system and can be calculated offline [7]. The quadratic linear regulator is now considered for an infinite duration interval subject to the following cost index [8] Equation (18) [3].

$$J_\infty = \frac{1}{2}\int_0^\infty \left(x^T(t)Qx(t) + u^T(t)Ru(t)\right)dt. \tag{18}$$

In this case since $t \to \infty$, the value of the matrix S(t) tends to its stationary value, and therefore, $S(t) = S$ and $\dot{S}(t) = 0$ [9]. The dynamic equation of Riccati, Equation (16), becomes an algebraic equation of Riccati that along with the static gain of Kalman remains Equation (19) and Equation (20) [10].

$$A^TS + SA - SBR^{-1}B^TS + Q = 0, \tag{19}$$

$$K = R^{-1}B^TS. \tag{20}$$

Therefore, the status feedback control will be Equation (21) [3].

$$u^*(t) = -R^{-1}B^TSx^*(t) = -Kx^*(t). \tag{21}$$

3. Application

In a DC motor with constant armature current (Figure 1) in which the resistance values (R) of the inductance (L) and the input voltage (Vin) are given, these values and parameters are given by circuit design. This motor is used to feed a power circuit that is used in most electrical systems [11].

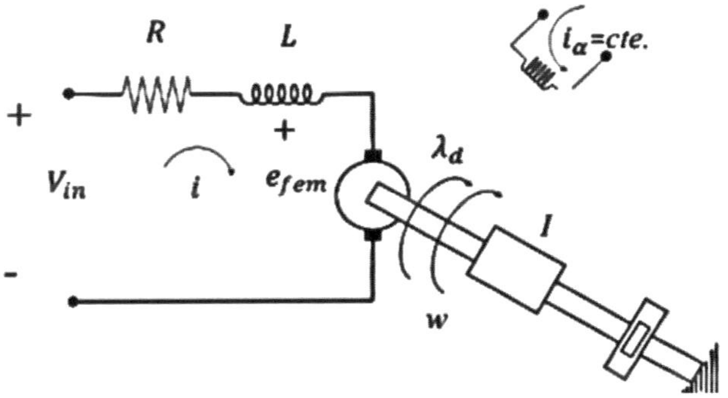

Figure 1. DC Motor.

The set of equations of state is given by Equation (22) and Equation (23) [3]. The cues represent the control variables in terms of the parameters and the input variables.

$$\frac{di(t)}{dt} = -\frac{R}{L}i(t) + \frac{1}{L}{v^2}_{in(t)}, \tag{22}$$

$$\frac{dw(t)}{dt} = \frac{K}{I}i(t). \tag{23}$$

where $\lambda_d(t) = Ki(t)$ is the instantaneous torque developed; $K = K_tI_a$ is the product of constant torque and armature current, and I is the moment of angular inertia the cost function to be minimized is Equations (24) [12].

$$J = \int_0^\infty [i^2(t) + w^2(t) + {v^2}_{in}]dt. \tag{24}$$

Journal of Physics: Conference Series **2153** (2022) 012011 doi:10.1088/1742-6596/2153/1/012011

4. Results

The frequency responses of the problem are shown below using Bode diagrams and the response of the system using a setpoint tracking and disturbance rejection diagram. The frequency response of the problem is shown in Figure 2; by means of Bode diagrams represented in Figure 2(a) and Figure 2(b), the frequency response of the system was characterized. In each of these Figures a magnitude diagram and a phase diagram are presented, observing in both cases, Figure 2(a), and Figure 2(b), that the magnitude stabilized as the frequency advances.

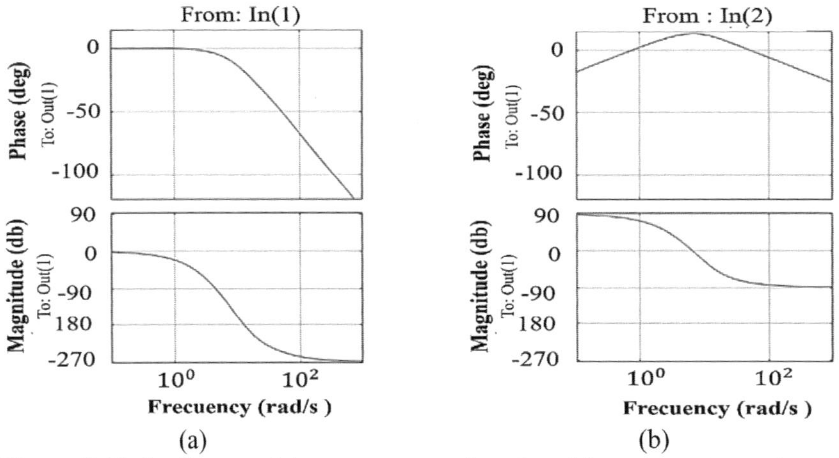

Figure 2. Bode diagram: (a) response to an impulse, (b) response to a unit step.

Figure 3 represents the response of the system. In Figure 3 (a) and Figure 3 (b) we observe the response of the system to an impulse and to a unit step respectively. In them the response of the system follows the reference even when there is a disturbance. A disturbance is physically a load torque that acts on the inertia of the motor. This charge torque acts as an additive term in the second equation of state that is divided by I, just like all other terms in this equation. These disturbances produce peaks in the control signal that by means of control return towards the reference as shown in Figure 3(a) and Figure 3(b).

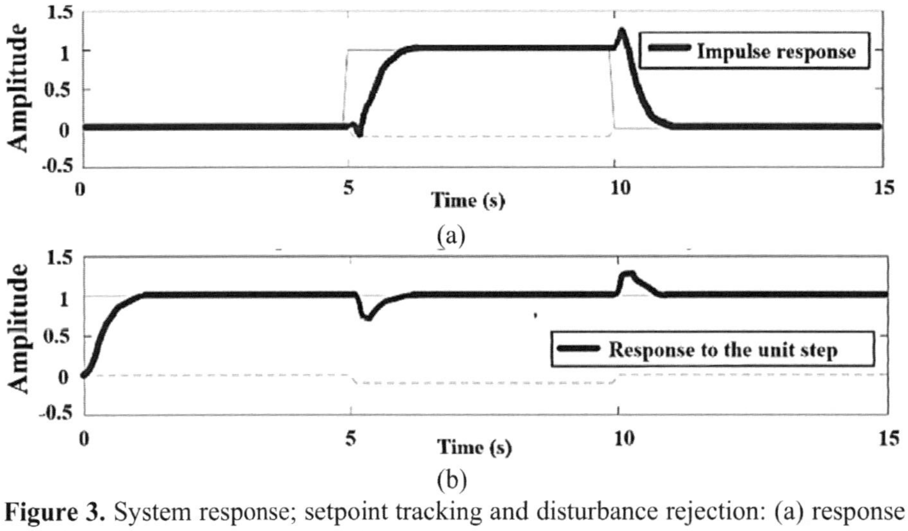

Figure 3. System response; setpoint tracking and disturbance rejection: (a) response to an impulse (b) response to a unit step.

5. Conclusions

It was demonstrated in the research that the control system based on a quadratic line regulator is effective due to its fast stability and damping times that bring the system to the proper reference. Power electronics systems based on many direct current motors need controllers that allow their operation when coupled to the power systems, for this purpose the control system is necessary to guarantee operation.

The approach of a control system based on functionalities, which characterize the dynamics of the system, is more efficient than doing it through windows over the entire operating trajectory. It is necessary that the motors that operate within a wide range of nominal operating points, be linearized so that they are designed at the operating points, outside of them efficiency begins to be lost.

The controller adaptation parameters are important since the rapid damping of the output signal depends on them, for this case it is proposed to implement the heurisks algorithms that allow obtaining fast results in terms of computational time. Depending on the selected weight matrices, different graphs are obtained that represent the feedback of the state as a function of the nominal operating points. The appropriate choice of weight matrices for the realization of a specific vector control scheme is still a problem. open research

References

[1] Kirk D E 2004 *Optimal Control Theory: An Introduction* (New York: Dover Publications)

[2] Athans M, Falb P 2007 *Optimal Control: An Introduction to the Theory and its Applications* (New York: Dover Publications)

[3] Naidu D S 2003 *Optimal Control Systems* (Boca Raton: CRC Press)

[4] Montserrat J T, Angel L A 2007 *Control Óptimo* (Valencia: Universitat Politècnica de València)

[5] Bayati M, Eshtehardiha S, Kiyoumarsi A, Ataei M 2007 Optimizing LQR and pole placement to control buck converter by genetic algorithm *International Conference on Control, Automation and Systems* (Seoul: IEEE

[6] Ruba M K, Hummadi A M 2012 Simulation of optimal speed control for a DC motor using linear quadratic regulator (LQR) *Journal of Engineering* **18(3)** 340

[7] Al Salameh Y A, Dalabeeh A S, Al Khalid K M 2019 Optimal state feedback controller for separately excited dc motor *International Journal of Computer Science and Mobile Computing* **8(6)** 187

[8] Obeid Ahmed A H O 2013 Optimal speed control for direct current motors using linear quadratic regulator *Journal of Science and Technology - Engineering and Computer Sciences* **14(2)** 48

[9] Moradi M, Ahmadi A, Abhari S 2010 Optimal control based feedback linearization for position control of DC motor *2nd International Conference on Advanced Computer Control* (Shenyang: IEEE)

[10] Murtaza G, Bhatti A I 2012 Control of DC motors using sliding mode *9th International Bhurban Conference on Applied Sciences & Technology (IBCAST)* (Islamabad: IEEE)

[11] Xu C, Li M, Pan F 2011 The system design and LQR control of a two-wheels self-balancing mobile robot *International Conference on Electrical and Control Engineering* (Yichang: IEEE)

[12] Abut T 2016 Modeling and optimal control of a DC motor *Int. J. Eng. Trends Technol* **32(3)** 146

Finite element analysis of the voltage induced by a pair of electrodes in a cell membrane

D F Devia[1], F Mesa[1], and D M Devia[1]
[1] Universidad Tecnológica de Pereira, Pereira, Colombia

E-mail: dfdevia@utp.edu.co, femesa@utp.edu.co

Abstract. This document analyses the distribution of the electric potential due to two electrodes that excite a membrane. The solution obtained numerically was approximated by means of the finite element method. The qualitative analysis of the results allows to know the intensity of the electric potential along the membrane in such a way that said electrodes that excite the membrane can be strategically located for academic and clinical purposes. Additionally, in a two-dimensional domain that represents the geometry of a membrane, the finite element method was executed, which allowed the behavior of the potential to be analyzed for an electric pulse at any point in the membrane. The above was generated by two electrodes.

1. Introduction

In general, differential equations are difficult to solve analytically due to the non-linearity of the mathematical models and the properties of the medium or materials involved [1]; by means of the finite element method and using the Galerkin approximation, it is possible to convert a differential equation that models a physical phenomenon in continuous time into an equation that can be solved in discrete time using test functions. Combinations of these test functions make it possible to approximate the answer to the differential equations by means of numerical analysis. Otherwise, the finite element method seeks the participation (contributions) of each of these test functions in such a way that the numerical solution approximates the analytical solution; the choice of test functions by means of the Galerkin approximation will consist of finding test functions that are close to the true (analytical) response [2].

In order for the differential equations to be solved, using the finite element method, the domain must be discretized into uniformly spaced fragments according to the domain, which means that they are linear one-dimensional elements, two-dimensional triangles or quadrilaterals or elements of 5 or 5. 8 tetrahedral nodes based on the precision that is required in three dimensions, so prior to the division of the domain into fragments, the requirement is created to find a set of linear equations that allow being built from the concepts listed above, a global function that characterizes the solution of the differential equation and this requires a directly proportional relationship between the number of linear coefficients and precision, which means calculating the coefficients that multiply each test function, and then perform a combination linear of these [3].

2. Previous mathematical concepts

Let's start from the following problem represented in Equation (1) [4].

$$w_{xx} + w_{yy} = 0, \tag{1}$$

with $(x,y) \in \Omega$ and with boundary conditions $w(x_0,y_0) = \alpha_1$, and $w(x,y) = h(x,y)$. In general, for $\Omega \subset \mathbb{R}^2$domains, $\Omega_1,\Omega_2,\Omega_3 \ldots,\Omega_m$ consecutive subdomains are fragmented into m triangular subdomains, thus constructing a mesh by means of triangles like the one in Figure 1 in such a way that on each triangular element we evaluate the conditions of border. Suppose that X^h is a space of polynomials of degree 1 defined in pieces $(\Omega_1,\Omega_2,\Omega_3 \ldots,\Omega_m)$, so X_m is required to be a space of $P_1(\Omega_m)$ and for this to be true a test function $\psi(x,y) = a + bx + cy$. The continuity of the functions defined in pieces is guaranteed since for this test function the values of the vertices (nodes) are known; the variable of interest at each node $(i \leftrightarrow 1, j \leftrightarrow 2, k \leftrightarrow 3)$takes the values $\emptyset_i, \emptyset_j, \emptyset_k$. It is easy to see that τ_1, τ_2, τ_3 can be obtained from the system of equations proposed in Equation (2), as shown in Equation (3) and Equation (4) [4].

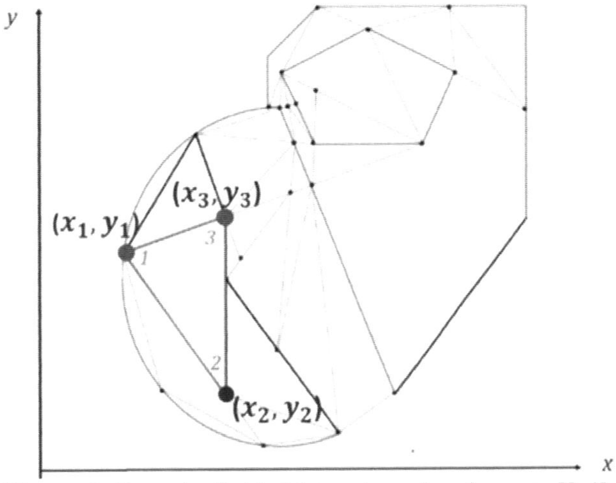

Figure 1. Domain divided into triangular elements [3,4].

$$\begin{bmatrix} \tau_1 + \tau_2 x_i + \tau_3 y_i \\ \tau_1 + \tau_2 x_j + \tau_3 y_j \\ \tau_1 + \tau_2 x_k + \tau_3 y_k \end{bmatrix} = \begin{bmatrix} \emptyset_i \\ \emptyset_j \\ \emptyset_k \end{bmatrix}, \tag{2}$$

$$\begin{bmatrix} (x_j y_k - x_k y_j)\emptyset_i + (x_k y_i - x_i y_k)\emptyset_j + (x_i y_j - x_j y_i)\emptyset_k \\ (y_j - y_k)\emptyset_k + (y_k - y_i)\emptyset_j + (y_i - y_j)\emptyset_k \\ (x_k - x_j)\emptyset_k + (x_i - x_k)\emptyset_j + (x_j - x_i)\emptyset_k \end{bmatrix} = Q^m \begin{bmatrix} \tau_1 \\ \tau_2 \\ \tau_3 \end{bmatrix}, \tag{3}$$

$$Q^m = \frac{1}{2}\left[(x_i y_j - x_j y_i) + (x_k y_i - x_i y_k) + (x_j y_k - x_k y_j)\right]. \tag{4}$$

The numerical solution can be written as shown in Equation (5), being a linear combination of polynomial functions of degree 1 of triangular form, Equation (5) [4].

$$\emptyset = P_i(x,y)\emptyset_i + P_j(x,y)\emptyset_j + P_k(x,y)\emptyset_k. \tag{5}$$

For this solution $\emptyset$ the partial derivatives are given by Equation (6) [4].

$$\begin{bmatrix} \frac{\partial \emptyset}{\partial x} \\ \frac{\partial \emptyset}{\partial y} \end{bmatrix} = \frac{1}{Q^m} \begin{bmatrix} (y_j - y_k) & (y_k - y_i) & (y_k - y_i) \\ (y_k - y_i) & (y_k - y_i) & (x_j - x_i) \end{bmatrix} \begin{bmatrix} \emptyset_i \\ \emptyset_j \\ \emptyset_k \end{bmatrix}. \tag{6}$$

Applying Galerkin to solve Equation (1) it is possible to find a functional that represents the numerical solution of the equation at each node, according to Equation (7) [5].

$$v_h(x,y) = \sum_{i=1}^{N} \gamma_i \emptyset_i(x,y) \in \Omega, \tag{7}$$

where γ_i are the coefficients of the test functions, $i = 1,2,3 \ldots N$. From the matrix $[\tau]$ of coefficients and the matrix $\vec{\beta}$ of boundary conditions, with the Galerkin approximation we obtain a linear system of equations of the form $[\tau]\vec{\gamma} = \vec{\beta}$ and the internal product between the interpolation functions (functional) is defined as the integral given in Equation (8) [5].

$$J\big(w(x,y)\big) = \iint_\Omega \left\{\frac{1}{2}\left[(w_x)^2 + \left(w_y\right)^2\right] + f(x,y)w(x,y)\right\} dxdy. \tag{8}$$

By approximating the behavior of the solution and given that $w(x,y) \longrightarrow v_h(x,y)$ is a numerical approximation of the functional [5] defined in pieces for each subdomain, we obtain Equation (9) and Equation (10) [5].

$$J\big(w(x,y)\big) = J\big(\sum_{i=1}^{N} \gamma_i \emptyset_i(x,y) \in \Omega\big), \tag{9}$$

$$J\big(\sum_{i=1}^{N} \gamma_i \emptyset_i \in \Omega\big) = \iint_\Omega \left\{\frac{1}{2}\left[\left(\sum_{i=1}^{N} \gamma_i \emptyset_{i_x}\right)^2 + \left(\sum_{i=1}^{N} \gamma_i \emptyset_{i_y}\right)^2\right] + f(x,y) \sum_{i=1}^{N} \gamma_i \emptyset_i\right\} dxdy. \tag{10}$$

However, if we intend to solve the equation $w_{xx} + w_{yy} = w_t$ which contains a term that shows the dynamics of the temporal variable then the solution will be given by Equation (11) [6-8].

$$v_h(x,y,t) = \sum_{i=1}^{L} W_i(t) \gamma_i \emptyset_i(x,y), \tag{11}$$

where $W_i(t)$ is a time-varying test function.

3. Application in the analysis of current in a cell membrane

By means of the circuit equivalent known as parallel conductance presented in Figure 2, it is possible to model the behavior of the electric current that circulates through a cell membrane when it is activated, as proposed in [8]. The advantage of this model is the separation of the leakage ions, sodium ions and potassium ions in branches [9], generating a current due to sodium ions, another due to potassium ions a current due to leakage ions and a displacement current, which occurs due to the uniformity of the cell membrane.

In Figure 2, corresponding to the equivalent circuit for the action of a membrane , the equivalent voltage V_m, the input and output conductances C_{m_1} and C_{m_2} are observed, corresponding to the energy stores in this circuit and the electrical conductance g_{N_a}, g_K,and g_L associated with the currents of sodium, potassium and leakage ions that correspond to the energy dissipaters in this circuit.

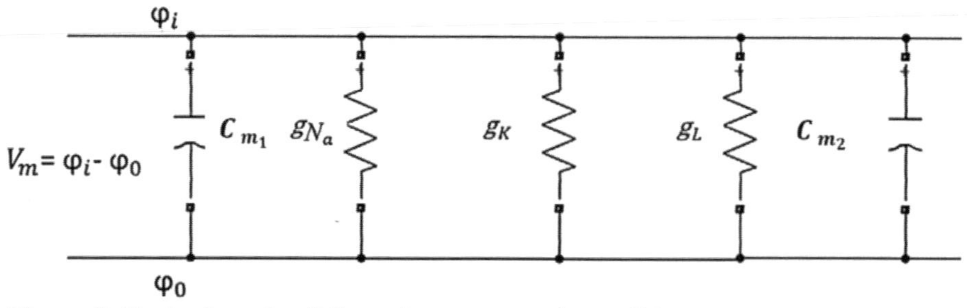

Figure 2. Equivalent circuit for action on a membrane [8].

The excitation of the membrane by means of an electrode is modeled using the laws of circuits and is expressed in Equation (12) [10], which allows us to determine the equivalent current (i_m).

$$C_{m_1}\frac{\partial V_m}{\partial t} + g_{N_a}(V_m - V_{N_a}) + g_K(V_m - V_K) + g_L(V_m - V_L) + C_{m_2}\frac{\partial V_m}{\partial t} - i_m = 0. \quad (12)$$

If a pair of constant sources excite the cell membrane, the dynamics produced by the source is negligible or null, consequently, $\frac{\partial V_m}{\partial t} = 0$. Therefore, and considering different experiments, it is necessary to feed said membrane with variable sources in time [11].

The propagation in the membrane due to the impulse must be modeled as the distribution of the potential across a membrane in a uniform medium with characteristics imposed by sodium, potassium, and other chemical combinations typical of the cell membrane. From the circuit shown in Figure 3 it is possible to obtain by means of Ohm's law a relationship between the current i_m, equivalent resistance (r_m), and V_m in Laplacian terms, Equation (13) [7,8].

$$i_m = r_m\left(\frac{\partial^2 V_m}{\partial x^2} + \frac{\partial^2 V_m}{\partial y^2}\right). \quad (13)$$

In this way the following partial derivative equation is obtained, Equation (14) [7,8], which is a parabolic differential equation [10].

$$r_m\left(\frac{\partial^2 V_m}{\partial x^2} + \frac{\partial^2 V_m}{\partial y^2}\right) = (C_{m_1} + C_{m_2})\frac{\partial V_m}{\partial t} + g_{N_a}(V_m - V_{N_a}) + g_K(V_m - V_k) + g_L(V_m - V_L). \quad (14)$$

Next, we show in the graph of Figure 3 the simplified model of current per unit length; in the model, the variable r_m represents the resistance per unit length, inside and outside the membrane, x and l correspond to the real part of the equivalent resistance (conductance) and the length of the circuit, respectively. This simplified circuit describes the behavior of the electrical variables in the membrane in terms of the second derivatives.

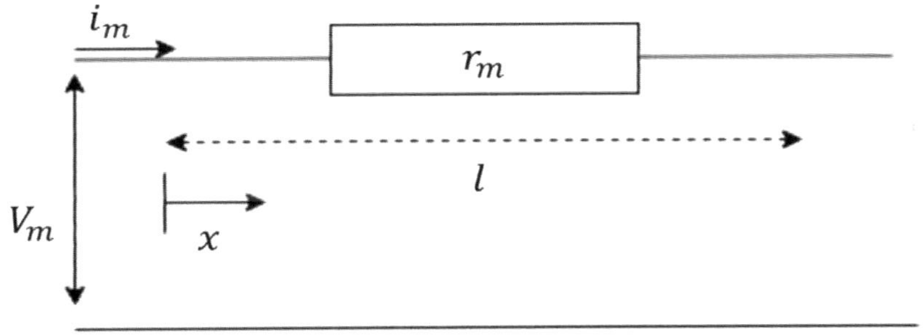

Figure 3. Simplified model dependent on the geometry of the membrane [2].

The solution of the partial differential equation obtained above is given by Equation (11), where the part that depends on time will take the form of the time-varying source in such a way that it excites the membrane and cell activity, Equation (15) [11].

$$W_i(t) = Aw(t - t_0). \quad (15)$$

The cell membrane characteristics for analysis in this paper are as shown in Table 1 [12], where $r_m = h/2\rho$, h is the height of the membrane (cm), a is the width of the membrane (cm), and ρ is the resistivity per longitudinal unit (Ω cm).

Table 1. Membrane characteristics.

Parameter	Value	Parameter	Value
r_m	$10/\Omega$	V_L	-54.4 mV
C_m	1.0 μF	a	1.0 cm
g_{Na}	1.0 μF	h	0.5 cm
g_K	36.0 mS	$W_i(t)$	$Aw(t-t_0)$
g_L	0.3 mS	A	$20000\left[\frac{\mu A}{cm^2}\right]$
V_{Na}	50.0 mV	t_0	0.5 ms
V_K	-77.0 mV		

4. Results

In Figure 4 we observe the distribution of Vm in the membrane; the maximum value of the voltage on the membrane is 1.25V. This value allows the membrane to be stimulated at the location of the electrodes, on the other hand, this potential value is attenuated as it passes through the chemical properties of the medium, reaching a value equal to zero (purple lines).

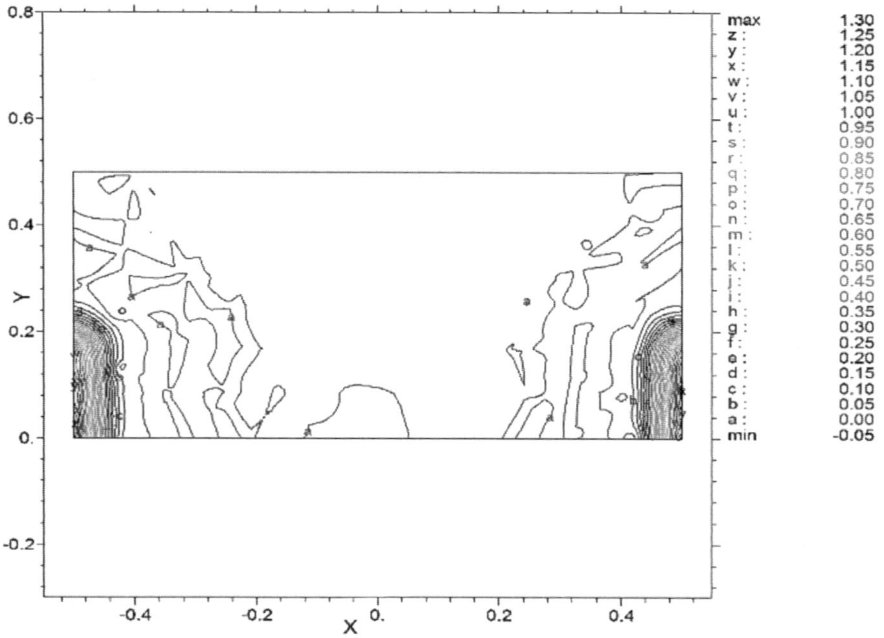

Figure 4. Distribution of the variable Vm in the membrane; X and Y are distance in nanometers (nm).

In Figure 5(a) it is possible to appreciate the distribution of the electric potential due to the pulse of two electrodes (X, Y Distance nanometers). The distribution shown in this figure refers to its cross section (through the thickness of the membrane), in this way the value of the potential due to the geometry of the membrane is known.

In Figure 5(b) the behavior in time T (in seconds) of the potential V_m (in volts) due to the pulses emitted by the two electrodes can be analyzed at the point (0,0) of the membrane. For sufficiently low levels of current density, the behavior of the current is capacitive through the admittance of the membrane. Otherwise, for high currents that exceed the capacity of the membrane, it would produce stress levels in the membrane that are too high and chemically irreversible effects.

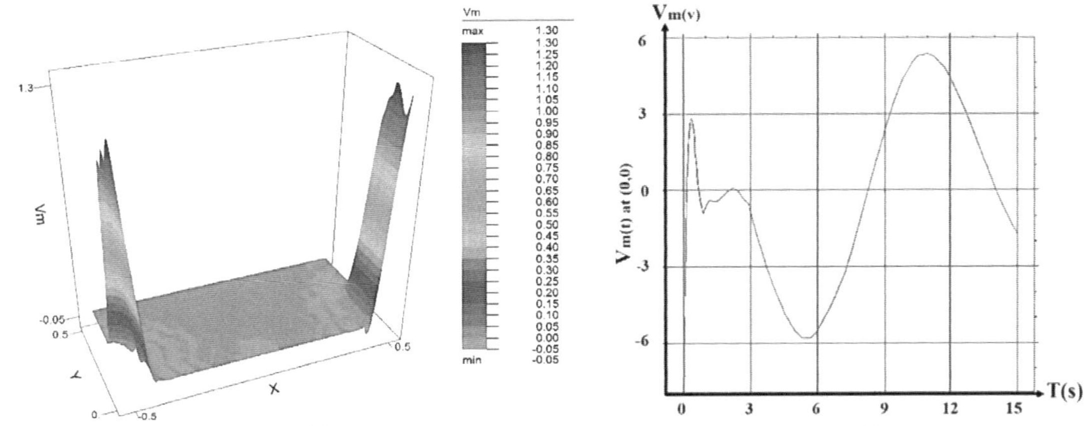

(a) (b)

Figure 5. (a) Arrangement of the two electrodes crossing the membrane and the values of the variation of the voltage in the membrane; (b) dynamics of the voltage in the membrane as a function of time.

5. Conclusions

Stimulation by means of two electrodes in tissues is produced by a time-varying magnetic field that results from the induced current flow through the membranes. Without such current flow, depolarization does not occur, and excitation cannot occur as presented in Figure 4.

In this paper it was analyzed for a single membrane under the excitation of two electrodes in a conductive medium given by the chemical components of the membrane, unfortunately, it is not possible to study the excitation of membranes in a general sense, but it is necessary to analyze the problem from the geometry, the structure, and the characteristics of the membrane. When the current density levels are very low, the behavior of the current is capacitive through the admittance of the membrane; however, for very high currents that exceed the capacity of the membrane, we observe stress levels in the membrane that are too high and chemically irreversible effects.

The stimulation electrodes must be applied to exclusively use their capacity and therefore, keep them within their operating range that under the current conditions at the electrode are not harmful to the membrane. The finite element method proves to be a numerical technique that can be used to solve many problems in physics, particularly those represented by parabolic differential equations.

References

[1] Reddy B D 1998 *Introductory-Functional-Analysis-With-Applications-to-Boundary-Value-Problems-and-Finite-Elements* (New York: Springer)
[2] Mesa F, Devia D M, Correa G 2020 Behavior of the voltage produced by an electrode in a membrane by means of the finite element method *Journal of Physics Conference Series* **1704(1)** 012004:1
[3] Pozrikidis C 2014 *Introduction to Finite and Spectral Element Methods Using MATLAB* (Boca Raton: CRC Press)
[4] Logan D L 2012 *A First Course in the Finite Element Method* (Stamford: Cengage Learning)
[5] Quarteroni A, Sacco R, Saleri F 2007 *Numerical Mathematics. Texts in Applied Mathematics* (Berlin: Springer)
[6] Sui, X, Huang Y, Chai X, Wang G 2012 3D finite element analysis of the electric field generated by epi-retinal MEMS electrodes *Biomedical Circuits and Systems Conference* (Hsinchu: IEEE)
[7] Howell B, Naik S, Grill W M 2014 *IEEE Transactions on Biomedical Engineering* **61(2)** 297
[8] Wtorek J 2000 *Annals of Biomedical Engineering* **28** 1352
[9] Kotnik T, Miklavčič D, Slivnik T 1998 *Bioelectrochemistry and Bioenergetics* **45(1)** 3
[10] Shagoshtasbi H, Lee Y K 2011 A new equivalent circuit model for micro electroporation systems *International Conference on Nano/Micro Engineered and Molecular Systems* (Kaohsiung: IEEE)
[11] Butson C R, McIntyre C C 2005 *Journal of Neural Engineering* **3(1)** 1
[12] Butson C R, Maks C B, McIntyre C C 2006 *Clinical Neurophysiology* **117(2)** 447

Simulation of inference test performance for minimum inhibitory concentration censored data

H A Torres-Mantilla[1], L Cuesta-Herrera[2], J E Andrades-Grassi[3], and G Bianchi[3]

[1] Universidad de Santander, Bucaramanga, Colombia
[2] Universidad Católica del Maule, Talca, Chile
[3] Universidad de Los Andes, Mérida, Venezuela

E-mail: h.a.torresmantilla@hotmail.com

Abstract. The estimation of the minimum inhibitory concentration is usually performed by a method of serial dilutions by a factor of 2, introducing the overestimation of antimicrobial efficacy, quantified by a simulation model that shows that the variability of the bias is higher for the standard deviation, being dependent on the metric distance to the values of the concentrations used. We use a methodological approach through modeling and simulation for the measurement error of physical variables with censored information, proposing a new inference method based on the calculation of the exact probability for the set of possible samples from n measurements that allows quantifying the p-value in one or two independent sample tests for the comparison of censored data means. Tests based on exact probability methods offer a reasonable solution for small sample sizes, with statistical power varying according to the hypothesis evaluated, providing insight into the limitations of censored data analysis and providing a tool for decision making in the diagnosis of antimicrobial efficacy.

1. Introduction

Since their discovery in the middle of the last century, antibiotics have been used to combat infectious diseases [1]. They have contributed to the increase in life expectancy by reducing mortality and morbidity, particularly in children. The emergence of resistant strains of bacteria with increasingly diverse antibiotic resistance mechanisms has been described with increasing frequency [1, 2]. The use of antibiotics effective against pathogenic bacterial strains contributes to disease control. The measurement of antimicrobial efficacy is usually performed by means of the minimum inhibitory concentration (MIC); by this method it can be established whether the concentration reached in blood is high enough to prevent bacterial growth [2–4].

Antimicrobial efficacy can be estimated from the sample obtained from affected host tissues using different methodologies [5, 6]. The data obtained may be categorical, as when limited to determining whether a bacterial strain is susceptible or resistant, or may even be continuous quantitative data depending on the estimation methodology. The most commonly used methodology involves discretization as the possible values are limited only to a set of pre-determined concentrations of the antibiotic substance.

The measurement of a characteristic of a continuous nature using methodologies that lead to loss of information by censoring leads to a detriment in the ability of estimates and hypothesis testing to approximate reality [7–9]. According to Fedorov, *et al.* [10], this loss of information, in terms of Fisher information for the case of dichotomization of Gaussian data, is close to 36% [5]. The loss of information is dependent on the discretization or censoring limits chosen and the metric distance from them to the unknown parameters. This loss of information leads to the need for a larger sample size to achieve the same statistical power [6, 10].

The analysis of antibiotic susceptibility data is highly dependent on the measurement methodology [5]; logistic regression models or other analyses for categorical data would be appropriate for performing inference tests or exploring causal relationships with susceptibility data expressed in dichotomous form. This could be useful when several sources of information are presented, e.g. laboratories with various susceptibility cut-off points. For the comparison of two antibiotic substances for the same strain, tests of independence could be used with a lost significance of statistical power, especially if control of covariates is required [5].

Annis and Craig [11] propose to correct for bias in the MIC estimation by an application of Aitchison's theory of remainders, assuming that the true value of MIC consists of a mixture of normal distributions and performing an imputation of missing values by the expected value of the logarithm of the likelihood function. In contrast Zhang, *et al.* [12] employs an adjustment based on a Bayesian hierarchical model requiring data with a low amount of censoring. Other data analyses for censored data, e.g. mixed models or accelerated failure time models, which involve assuming the metric properties of the measurements while retaining more information, may not be suitable if there is a high degree of censoring or when the concentrations used in the measurements cannot be adjusted, for example when the distribution of antibiotic susceptibility is not sufficiently studied [5, 13].

2. Methodology

To describe susceptibility to an antibiotic we need to estimate the value of μ. It is reasonable to assume that the value of σ is proportional to the value of μ. In this way, if a strain of a bacterial species has a high μ value of X (that is, it is inhibited only at high concentrations of the antibiotic), the variability of X will be higher than that of the variable X for a strain of a bacterial species with a low μ value. For example, if the average X for strain A is 50 and the average X for strain B is 10, the variability of X_A is expected to be higher; for the proportion, it can be assumed that σ is proportional to μ (constant coefficient of variation) or that σ is proportional to the square of μ (coefficient of variation proportional to the mean) as possibilities.

Initially, the existence of bias in the estimation of the mean and standard deviation was verified according to the censoring intervals conditioned by the dilution method used for the measurement of the MIC. Starting with simulated data from an independent normal process, the data were categorized according to the censoring intervals corresponding to the successive dilutions in a factor of 2. We considered obtaining a set of 10000 samples of 6 and 10 measurements for which the mean and standard deviation were calculated for the censored data; by direct calculation of the bias and in analyses for each sequence of dilutions, the variation of the bias with respect to the actual values of the true mean MIC for the sample mean and sample standard deviation was studied. Subsequently, two sets of 5000 independent samples of 6 and 10 measurements were simulated by assuming equal means for the two sets of actual MIC values.

In order to perform the comparison between the two samples, the use of a methodology to calculate the exact probability based on the assumption of a normal distribution for the actual MIC values is proposed. The independence of the observations for each set of measurements, the normality of the underlying distribution of the actual MIC and the constant proportionality between the mean and the standard deviation were used as assumptions. The set of all possible sampling arrangements for one sample and for comparison of two samples were computed recursively.

The exact probability was computed using the normal distribution and the cumulative probabilities for the censored values using these assumptions for four consecutive dilutions, assuming also the very low probability ($< 10^3$) that a sample of 10 measurements or less would show data with a range of values covering more than 4 dilutions for the same population. The sum of the probabilities of sampling arrangements whose mean difference of is equal to or greater than the mean difference under evaluation is considered for the calculation of the p-value.

The t-test and the proposed test for the comparison of means between two substances with a constant difference between population means were applied by changing the metric distance between the population means and the censoring limits; for the computation of the p value, a null difference between population means is used, while for the computation of the power of the tests, progressively larger differences between population means were used. Concentrations are expressed in absolute terms (in ug/ml) as well as in terms relative (in percent) to the next higher censoring limit. Using the relative values allows extrapolation of the results to other concentrations because the results are reproduced at lower or higher concentrations as long as the percentage distance from the censoring limits is maintained

For the calculation of the power each test applies the inference test for a progressively larger mean difference for the same significance level. The sum of the probabilities of obtaining a censored mean difference greater than or equal to the critical value according to the chosen significance level corresponds to the power of the test. Simulations were performed for 10000 samples of 10 measurements and 10000 samples of 6 measurements to verify the consistency of the results and to obtain the graphical results; R software version 4.1.1.1 were used [14].

3. Results

In Figures 1(a) and Figure 1(b) show that there are variability in the bias in the estimation of the arithmetic mean. The variability of the bias is greater for the standard deviation that even changes sign, being dependent on the metric distance to the values of the concentrations used. This leads to an overestimation or underestimation of the MIC variability which would condition differences in the statistical power of the tests depending on the proximity of the mean of the actual MIC distribution with respect to the censoring bounds.

For the mean difference tests, the censored interval is assumed as the hypothetical mean, being the value with the lowest statistical power among those evaluated. For a sample of 10 measurements, with a significance level of 10%, that a mean difference greater than or equal to 45% about the mean is statistically significant ($p = 0.0841$). Using as a hypothesis the equality of means corresponding to a censoring interval of 50 ug/ml, corresponds to a difference of censored means of 22.5 ug/ml. On the other hand, if a sample of 6 measurements is considered, a difference of means greater than or equal to 58% is considered statistically significant ($p = 0.0719$). For an assumption of means equal to 50 ug/ml corresponds to a difference greater than or equal to 29 ug/ml.

Table 1 and Table 2 show the results of the tests based on the proposed methodology according to the percentage and absolute difference between the observed means of censored data. In Table 1, for the case of 6 samples, it can be seen that the difference is significant, for a level of 5%, from a difference of observed means greater than 54% of the upper limit, which for the case of 50 ug/ml corresponds to a difference of means equals 27 ug/ml. On the other hand, Table 2, for the case of 10 samples, shows that it is significant with differences greater than 43% of the upper limit, which for the case of 50 ug/ml, corresponds to 21.5 ug/ml.

When evaluating the proportion of type 1 errors estimated by simulation for mean comparison tests for independent samples, with a significance level of 10% using both the t-test and the proposed test, using a hypothetical mean of 50 ug/ml (coinciding with a censoring limit), it was found that the t-test commits type 1 errors less than 3% of the time, while the proposed test commits type 1 errors 5% of the time. On the other hand, when 37.5 ug/ml (mean value between censoring limits) is used as the hypothetical mean, the frequency of type 1 error increases to 27% using the t-test while with the proposed test the frequency rises to 18%.

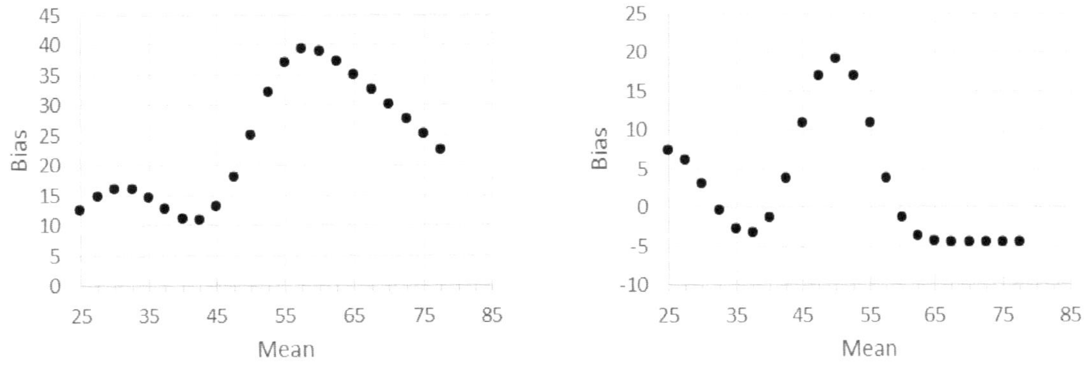

(a) (b)

Figure 1. Magnitude of the bias on the estimate of the statistics; (a) mean, (b) standard deviation.

Table 1. Test results for 6 samples of MIC by observed difference.

Percent difference about mean	Censored mean differences	p-value
17%	8.33	0.774
21%	10.42	0.387
25%	12.50	0.387
29%	14.58	0.387
33%	16.67	0.387
38%	18.75	0.146
42%	20.83	0.146
46%	22.92	0.146
50%	25.00	0.146
54%	27.08	0.038
58%	29.17	0.038
62%	31.25	0.038
67%	33.33	0.038
71%	35.42	0.006
75%	37.50	0.006
79%	39.58	0.006
83%	41.67	0.006

The power of the test is not easy to establish since the alternative hypothesis of the existence of a non-zero difference between the average varies according to the numerical distance between the averages and the censured measurements conditioned by the methodology of the experiment. In the case of the proposed exact test, for a sample of 10 measurements for the difference mentioned as significant of 45%, there is a power between 33% and 98% with an average of 73% to correctly reject the false null hypothesis.

In contrast, for a sample of 6 measurements for a difference of 58%, the power to correctly reject the null hypothesis is between 6% and 94% with an average value of 42%. In the case of t-tests, for samples of 10 measurements the power variation varies between 12% and 98%

depending on the censoring limits, with an average value of 46%. For samples of 6 measurements the power variation varies between 4% and 92% with an average value of 30%.

Table 2. Test results for 10 samples of MIC by observed difference.

Percent difference about mean	Censored mean differences	p-value
15%	7.50	0.503
18%	8.75	0.503
20%	10.00	0.503
23%	11.25	0.263
25%	12.50	0.263
28%	13.75	0.263
30%	15.00	0.263
33%	16.25	0.115
35%	17.50	0.115
38%	8.75	0.115
40%	20.00	0.115
43%	21.25	0.041
45%	22.50	0.041
48%	23.75	0.041
50%	25.00	0.041
53%	26.25	0.011
55%	27.50	0.011
58%	28.75	0.011
60%	30.00	0.011
63%	31.25	0.002

4. Conclusions

The bias in the estimation of the true mean minimum inhibitory concentration using the censured values varies according to the methodology generating the different concentrations and varies according to the metric distance of the true mean and the values of the concentrations used (censuring bounds). For the case of the actual standard deviation there is even more variability in the bias dependent on the metric distance to the concentration values used. The use of parametric methods is not advised even in cases where the standard deviation of the minimum inhibitory concentration is known in advance. An improvement in statistical power is observed when using the exact probabilistic method applied on the frequency distribution of observed values instead of assuming the unbiased estimate of the standard deviation assumed by the parametric test. Tests based on exact probability methods offer a reasonable solution for the case of small samples, with statistical power varying widely depending on the hypothesis evaluated.

References

[1] Bérdy J 2012 Thoughts and facts about antibiotics: where we are now and where we are heading *The Journal of antibiotics* **65(8)** 385

[2] Humphries R M, *et al.* 2018 CLSI methods development and standardization working group best practices for evaluation of antimicrobial susceptibility tests *Journal of Clinical Microbiology* **56(4)** e01934-17

[3] Mouton J W 2002 Breakpoints: current practice and future perspectives *International Journal of Antimicrobial Agents* **19(4)** 323

[4] Naylor N R, *et al.* 2018 Estimating the burden of antimicrobial resistance: a systematic literature review *Antimicrobial Resistance & Infection Control* **7(1)** 1

[5] Michael A, Kelman T, Pitesky M 2020 Overview of quantitative methodologies to understand antimicrobial resistance via minimum inhibitory concentration *Animals* **10(8)** 1405

[6] Mazloom R, Jaberi-Douraki M, Comer J R, Volkova V 2018 Potential information loss due to categorization of minimum inhibitory concentration frequency distributions *Foodborne pathogens and disease* **15(1)** 44

[7] Van de Kassteele J, van Santen-Verheuvel M G, Koedijk F D, van Dam A P, van der Sande M A, de Neeling A J 2012 New statistical technique for analyzing MIC-based susceptibility data *Antimicrobial Agents and Chemotherapy* **56(3)** 1557

[8] Turnidge J, Bordash G 2007 Statistical methods for establishing quality control ranges for antibacterial agents in clinical and laboratory standards institute susceptibility testing *Antimicrobial Agents and Chemotherapy* **51(7)** 2483

[9] Turnidge J, Kahlmeter G, Kronvall G 2006 Statistical characterisation of bacterial wild-type MIC value distributions and the determination of epidemiological cut-off values *Clinical Microbiology and Infection* **12(5)** 418

[10] Fedorov V, Mannino F, Zhang R 2009 Consequences of dichotomization *Pharmaceutical Statistics: The Journal of Applied Statistics in the Pharmaceutical Industry* **8(1)** 50

[11] Annis D H, Craig B A 2005 Statistical properties and inference of the antimicrobial MIC test *Statistics in Medicine* **24** 3631

[12] Zhang M, Wang C, O'Connor A 2020 A hierarchical Bayesian latent class mixture model with censorship for detection of linear temporal changes in antibiotic resistance *PloS one* **15(1)** e0220427

[13] Follmann D, Brittain E, Powers J H 2013 Discordant minimum inhibitory concentration analysis: a new path to licensure for anti-infective drugs *Clinical Trials* **10(6)** 876

[14] R Core Team 2020 *R: A Language and Environment for Statistical Computing* (Austria: R Foundation for Statistical Computing)

Classification of Parkinson's disease patients based on spectrogram using local binary pattern descriptors

E Gelvez-Almeida[1,2], A Vásquez-Coronel[1], R Guatelli[3], V Aubin[3], and M Mora[1]

[1] Laboratorio de Investigaciones Tecnologicas en Reconocimiento de Patrones, Universidad Católica del Maule, Talca, Chile
[2] Facultad de Ciencias Básicas y Biomédicas, Universidad Simón Bolívar, San José de Cúcuta, Colombia
[3] Departamento de Ingeniería e Investigaciones Tecnológicas, Universidad Nacional de La Matanza, Buenos Aires, Argentina

E-mail: e.gelvez@unisimonbolivar.edu.co

Abstract. Extreme learning machine is an algorithm that has shown a good performance facing classification and regression problems. It has gained great acceptance by the scientific community due to the simplicity of the model and its sola great generalization capacity. This work proposes the use of extreme learning machine neural networks to carry out the classification between Parkinson's disease patients and healthy individuals. The descriptor used corresponds to the feature vector generated applying the local binary Pattern algorithm to the grayscale spectrograms. The spectrograms are obtained from the audio signal samples from the considered repository. Experiments are conducted with single hidden layer and multilayer extreme learning machine networks comparing the results of each structure. Results show that hierarchical extreme learning machine with three hidden layers has a better general performance over multilayer extreme learning machine networks and a single hidden layer extreme learning machine. The rate of success obtained is within the ranges presented in the literature. However, the hierarchical network training time is considerably faster compared to multilayer networks of three or two hidden layers.

1. Introduction

Parkinson's disease (PD) is the second most common neurodegenerative disease after Alzheimer's disease. It involves slow and irreversible progression, the first symptoms becoming evident after several years of suffering from the disease. Motor symptoms, such as tremors, are the most popular ones. However, other symptoms can be, in many cases, incapacitating [1]. These are cognitive disorders, limb rigidity, gait and balance problems, and slowness. PD symptoms appear early as speech and voice disorders.

Several studies have proposed the processing of voice signals to obtain acoustic parameters as an objective and non-invasive method for PD detection [2]. Sakar B E, *et al.* [3] found that sustained vowels carry enough PD-discriminative information using machine learning tools. In Wodzinski M [4], a previously trained ResNet architecture is used for PD detection using ImageNet and singular value decomposition (SVD) databases. For classification, audio spectrograms of vowels with sustained phonation are used from the PC-GITA database [5]. The validation set has shown an accuracy rate higher than 90%. In Zahid, *et al.* [6], deep learning methodologies are combined with machine learning methodologies to classify Parkinson's disease patients. The multilayer perceptron applied to monologue recordings showed the highest accuracy of 99.7%. In Trinh, *et al.* [7], the detection of voice pathologies in patients who suffer from Parkinson's or Alzheimer's disease is proposed. A convolutional neural network (CNN) architecture is used with a classification accuracy higher than 95%.

The aforementioned works use machine learning methods known for their higher levels of fit to classification issues, reporting accuracy rates between 80% and 90%. It is well known that said methodologies experience a high computational cost in training algorithms and the estimation of the hyperparameters of the model. extreme learning machine (ELM) [8] has gained great acceptance from the scientific community because of the simplicity of the models and their generalization capacity. They began to be massively used to address complex problems of pattern recognition due to the low computational cost of their training algorithm.

In this paper, the use of ELM neural networks to address the classification issue of Parkinson's disease patients and healthy individuals is proposed. Experiments are conducted with single and multilayer ELM networks and results are compared. Grayscale spectrograms obtained from the voice signals from the repository presented in Giuliano M, *et al.* [9] are used. Descriptors for those images are generated using the local binary pattern (LBP) algorithm. Said algorithm highlights textures creating a feature vector. Multilayer extreme learning machine (ML-ELM) and hierarchical extreme learning machine (H-ELM) have shown a better performance addressing the classification problem of Parkinson's disease than single-layer ELM networks, with reduced training times.

2. Background

In this section, a brief description is presented of the spectrogram images, the LBP algorithm and the ELM variants used in this paper.

2.1. Spectrogram

A spectrogram is the representation of a signal in time. The result is a three-dimensional graphic of the spectrum of a signal formed by time, frequency, and amplitude of energy distribution. The amplitude of energy distribution is represented by color intensity. That is why it can be represented in a two-dimensional image, despite it having three dimensions. Speech representations through spectrograms have proven to be stable and robust, even with high noise levels [10, 11].

2.2. Local binary patterns

LBP is an operator for image texture analysis which has been used in several applications, including visual inspection, image retrieval, remote sensing, biomedical image analysis, face image analysis, motion analysis, environment modeling, and outdoor scene analysis libre [12]. The operator consists in labelling the pixels of an image by thresholding the neighborhood of each pixel and considers the result as a binary number. This operator traces the whole image and is mathematically represented as shown in Equation (1).

$$LBP_{(gp_x, gp_y)} \sum_{p=0}^{P-1} S(gp - gc) \times 2^P. \tag{1}$$

In Equation (1), gc is the intensity value of the central pixel, gp is the intensity of the neighboring pixel with index p and P is the number of sampling points on a circle of radius R

(circular neighborhood) [12]. Function S can be expressed as shown in Equation (2).

$$S_x = \begin{cases} 1 & \text{si} \quad x \geq 0 \\ 0 & \text{si} \quad x < 0' \end{cases} \tag{2}$$

This operator highlights textures from the original image creating a new one. The histogram is extracted from this new image, obtaining a feature vector as a result. This descriptor is used in the experiments to observe the classification behavior using the ELM, ML-ELM and H-ELM algorithms.

2.3. Extreme learning machine

Being $\aleph = \{(\boldsymbol{x}_i, \boldsymbol{t}_i) \in \mathbb{R}^n \times \mathbb{R}^m | i = 1, \ldots, N\}$ a random training set, $g(s) : \mathbb{R}^n \to \mathbb{R}^m$ an activation function, and L the number of neurons in the hidden layer with $L < N$ [13], the SLFN training algorithm is given by Equation (3).

$$\sum_{i=1}^{L} \boldsymbol{\beta}_i g(\boldsymbol{w}_i \boldsymbol{x}_j + b_i) = \boldsymbol{t}_j, \qquad j = 1, \ldots, N. \tag{3}$$

In Equation (3), $\boldsymbol{\beta}_i$ are the weights of the output layer, $\boldsymbol{w}_i$ are the weights of the hidden layer and b_i, the i-th bias of the hidden layer [8, 13]. $\boldsymbol{w}_i$ weights and b_i biases of the hidden layer are pseudo-randomly assigned. Equation (3) can be written in the matrix form $\boldsymbol{H\beta} = \boldsymbol{T}$, where, the $\boldsymbol{H}$ matrix in Equation (4) is called the matrix of the output hidden layer of the neural network [14]. Last, the $\boldsymbol{\beta}$ weights of the output layer are calculated using Equation (5).

$$\boldsymbol{H} = \begin{bmatrix} g(\boldsymbol{w}_1 \boldsymbol{x}_1 + b_1) & \cdots & g(\boldsymbol{w}_L \boldsymbol{x}_1 + b_L) \\ \vdots & \ddots & \vdots \\ g(\boldsymbol{w}_1 \boldsymbol{x}_N + b_1) & \cdots & g(\boldsymbol{w}_L \boldsymbol{x}_N + b_L) \end{bmatrix}, \quad \boldsymbol{\beta} = \begin{bmatrix} \boldsymbol{\beta}_1^T \\ \vdots \\ \boldsymbol{\beta}_L^T \end{bmatrix}, \quad T = \begin{bmatrix} \boldsymbol{t}_1^T \\ \vdots \\ \boldsymbol{t}_N^T \end{bmatrix}. \tag{4}$$

$$\boldsymbol{\beta} = \boldsymbol{H}^{\dagger} \boldsymbol{T}. \tag{5}$$

The $\boldsymbol{H}^{\dagger}$ matrix in Equation (5) is the Moore–Penrose generalized inverse of the $\boldsymbol{H}$ matrix [15].

2.4. Extreme learning machine auto encoder

Extreme learning machine auto encoder (ELM-AE) is an ELM modification to perform unsupervised learning. In this case, input data $\boldsymbol{x}$ is used as output data, so $\boldsymbol{t} = \boldsymbol{x}$. Weights and biases of the hidden layer are chosen to be randomly orthogonal in order to obtain a better performance [16]. In this sense, random orthogonal weights and biases are presented through Equation (6).

$$\boldsymbol{h}_i = g(\boldsymbol{w}_i \boldsymbol{x}_j + b_i), \quad \boldsymbol{w}^T \boldsymbol{w} = \boldsymbol{I}, \quad \boldsymbol{b}^T \boldsymbol{b} = 1. \tag{6}$$

In Equation (6), $\boldsymbol{w}$ and $\boldsymbol{b}$ are the weights and biases of the hidden layer, respectively [16]. Last, weights of the output layer are determined analytically with Equation (7).

$$\boldsymbol{\beta} = \left(\frac{\boldsymbol{I}}{C} + \boldsymbol{H}^T \boldsymbol{H} \right)^{-1} \boldsymbol{H}^T \boldsymbol{X}. \tag{7}$$

In Equation (7), $\boldsymbol{H}$ is the hidden layer output matrix, $\boldsymbol{X}$ is ELM-AE input and output data and C is a regularization parameter [16].

2.5. Multilayer extreme learning machine

ML-ELM is a multilayer neural network based on ELM-AE similar to deep networks and with better performance [16]. ML-ELM hidden layer weights are initialized with ELM-AE, without requiring fine tuning. In this case, $\boldsymbol{H}$ output matrix values of each hidden layer are obtained through Equation (8).

$$\boldsymbol{H}^k = g((\boldsymbol{\beta}^k)^T \boldsymbol{H}^{k-1}). \tag{8}$$

The $\boldsymbol{H}^k$ in Equation (8) is the k-th hidden layer output matrix [16]. Weights of the first hidden layer are the $\boldsymbol{\beta}^1$ results of the first ELM-AE that has samples $\boldsymbol{X}$ as input data. The weights that connect the different hidden layers are the $\boldsymbol{\beta}^{i+1}$ results of the ELM-AE that has the output results of the previous ML-ELM hidden layer as input [16]. Last, weights of the ML-ELM output layer are determined analytically using the Moore–Penrose generalized inverse [15].

2.6. Hierarchical extreme learning machine

Faced with the need to address application issues of greater complexity among the scientific community, an H-ELM, non-iterative multilayer perceptron model is proposed in en [17]. H-ELM training is divided into an unsupervised feature learning stage and a final supervised stage. In the first stage, an ELM-based sparse autoencoder (ELM-SAE) is introduced in the deep layers for high level feature extraction of input data. In ELM-SAE, weights and biases of the output layer are pseudo-randomly assigned, whereas the output layer weights matrix is subject to an optimization problem, with L1 norm regularization term, as shown in Equation (9).

$$\min_{\boldsymbol{\beta}} : F(\boldsymbol{\beta}) = \|\boldsymbol{H}\boldsymbol{\beta} - \boldsymbol{X}\|^2 + \lambda\|\boldsymbol{\beta}\|_1. \tag{9}$$

In Equation (9), λ is a regularization parameter and $\|\boldsymbol{\beta}\|_1$ represents the L_1 norm [17]. The fast iterative shrinkage-thresholding algorithm (FISTA) [18] is adopted to solve the convex scheme of Equation (9) and to further improve the training speed. Equation (10) establishes the connection between ELM-SAE and the H-ELM deep layers.

$$\boldsymbol{H}^{(k)} = g(\boldsymbol{H}^{(k-1)}\boldsymbol{\beta}^{(k)}). \tag{10}$$

In Equation (10), $\boldsymbol{H}^{(k)}$ is the k-th output layer matrix, $\boldsymbol{H}^{(k-1)}$ the $(k-1)$-th output layer matrix, $\boldsymbol{\beta}^{(k)}$ the k-th ELM-SAE output layer weights, and $g(\boldsymbol{H}^{(k-1)}\boldsymbol{\beta}^{(k)})$ is an activation function [17]. In the supervised stage, the final decision of the classifier is the consequence of the ELM method, which takes as input data the last hidden layer output matrix and the labels as output.

3. Materials and method

In order to conduct the experiments, a machine with central processing unit (CPU) intel(R) Core(TM) i7-8550U 1.80GHz-1.99GHz was used, with 4 cores and 8 logic processors, apart from 12 Gb RAM memory. Codes were made in MATLAB R2017b. Datasets and methods used in this study are presented in the following section.

3.1. Database description

A database whit grayscale spectrograms generated from audio signals from Parkinson's disease patients and healthy individuals were used for the experiments [9]. The database is composed of spectrograms images from 58 Parkinson's disease patients and 77 healthy individuals. Patients' ages vary from 38 to 79 years, and they have been diagnosed with Parkinson's for six years, on average.

3.2. Methodology of the experiments

Experiments are divided into two phases: a pre-processing phase of the database with the LBP algorithm and a classification phase with the ELM algorithm and the variants presented in this paper. The two phases conducted in this work are presented below.

3.2.1. Pre-processing phase In this phase, the LBP algorithm is applied to the spectrogram images to extract their features. In order to do so, the MATLAB `extractLBPFeature` function is used. The result obtained by this algorithm generated a feature vector of 59 values. A label is assigned to each value, 0 for healthy individuals and 1 for Parkinson's disease patients. After the features are extracted from the spectrograms, the database is distributed as shown in Table 1.

The training, validation and test sets are randomly selected to ensure confidence in the results and prevent its unintentional manipulation.

Table 1. Distribution of the database of spectrograms from Parkinson's disease patients and healthy subjects.

Patient	Training (60%)	Validation (20%)	Test (20%)	Total
Sick	34	12	12	58
Healthy	47	15	15	77

3.2.2. Classification phase The database described in the previous section is used for this phase; for the purpose of obtaining the optimum values of hyperparameters, different iterations are conducted in the experiments, varying the amount of neurons in the hidden layer from 100 to 5000 and the regularization parameter from 10^{-20} and 10^{20}. Moreover, the internal structure of the network varies from one hidden layer (standard ELM) to three hidden layers (ML-ELM and H-ELM). The sigmoid function of Equation (11) is used as an activation function for all cases.

$$g(w_i x_j + b_i) = \frac{1}{1 + e^{-(w_i x_j + b_i)}}. \quad (11)$$

In Equation (11), w_i and b_i are the weights and biases of the hidden layer, respectively, and x_j the input data to the network [8, 13]. Once the best results are obtained, a test set is carried out to verify the robustness of the network.

4. Results

To obtain hyperparameters presented in Table 2, tests were carried out with different combinations of the number of neurons in the hidden layer and the regularization parameters. Regularization hyperparameters improve accuracy, especially in those situations when data is imbalanced. The results of these experiments have shown that for this particular dataset, regularization parameter C does not improve results for the single-hidden-layer ELM network. That is why said parameter is not used.

Table 2. Hyperparameters used with best results in ELM and its variants.

Model	Hidden layer	Neurons	Regularization
H-ELM	3	$L_1 = L_2 = 500$, $L_3 = 1900$	$C = 10^{-4}$, $\lambda = 10^{-3}$
ML-ELM	3	$L_1 = L_2 = 1100$, $L_3 = 2700$	$C_1 = C_2 = 10^5$, $C_3 = 10^3$
ML-ELM	2	$L_1 = 1500, L_2 = 1200$	$C_1 = 10^5$, $C_2 = 10^6$, $C_3 = 10^2$
ELM	1	$L = 2100$	Not used

Once the hyperparameters are obtained, 10 iterations are made over the same dataset, registering accuracy average and standard deviation. The standard deviation among the results obtained varies from 0 to 0.009, which proves that the results are quite stable. Table 3 shows the comparison in terms of the accuracy and training speed between the standard ELM, ML-ELM, and H-ELM networks whit two and three hidden layers. The results presented were obtained using regularization parameters of Equation (7) and Equation (9).

As it can be seen in Table 3, multilayer networks (ML-ELM and H-ELM) have better validation and test results than a single-layer ELM network. For multilayer networks, the accuracy is 92.59% for validation and 81.48% for test. Said results are higher than the 88.89% and 77.78% obtained in validation and test, respectively, for single-layer ELM networks. Results for ML-ELM and H-ELM networks are equal for the training, validation, and test sets. However, training of H-ELM networks is performed 10.2 and 5.5 times faster in relation to three and two hidden layers of ML-ELM networks, respectively. It represents a significantly high and favorable reduction in training time which improves with databases at a larger scale.

Table 3. Accuracy in the classification of Parkinson's patients using standard ELM, ML-ELM, and H-ELM.

Model	Hidden layers	Training		Validation		Test	
		Accuracy	Time (s)	Accuracy	Time (s)	Accuracy	Time (s)
H-ELM	3	0.9012	0.2438	0.9259	0.0083	0.8148	0.0090
ML-ELM	3	0.9012	2.4849	0.9259	0.0055	0.8148	0.0060
ML-ELM	2	0.9012	1.3315	0.9259	0.0043	0.8148	0.0051
ELM	1	0.9136	0.0641	0.8889	0.0172	0.7778	0.0156

5. Conclusions

Standard extreme learning machine, multilayer extreme learning machine and hierarchical extreme learning machine to address the classification problem are used; said classification consists in identifying Parkinson's disease patients using a database composed of 135 spectrograms. The spectrograms are processed by the local binary pattern operator to extract features from the image to train the network; the dataset has been divided into three subsets, 60% for training (81 samples), 20% for test (27 samples), and 20% for validation (27 samples).

Results show that multilayer networks have a better performance than single-layer extreme learning machine networks. Multilayer extreme learning machine and hierarchical extreme learning machine networks present higher accuracy rates for the validation and test sets when compared to single-layer extreme learning machine networks. Although multilayer extreme learning machine and hierarchical extreme learning machine networks have the same accuracy rates for the three sets, training time with hierarchical extreme learning machine is 10.2 and 5.5 times faster than for three and two hidden layers multilayer extreme learning machine networks, respectively. Accuracy with multilayer extreme hierarchical extreme learning machine networks is 90.12% for the training set, 92.59% for the validation set, and 81.48% for the test set.

The hierarchical extreme learning machine structure which shows the best performance is the one formed by three hidden layers. The hyperparameters used $L_1 = L_2 = 1100$, $L_3 = 2700$, $C = 10^{-4}$, and $\lambda = 10^{-3}$, where L_1, L_2, and L_3 is the number of neurons in the first, second and third hidden layers, respectively, C and λ, the regularization parameter associated to the Moore–Penrose generalized inverse and fast iterative shrinkage-thresholding algorithm algorithm, respectively. The results show that the local binary pattern operator is a good alternative to generate spectrogram image descriptors to classify through hierarchical extreme learning machine networks. In future works, it is intended to try local binary pattern descriptors with other databases of larger-scale images using different extreme learning machine variants to compare accuracy and training times.

Acknowledgments
This work was funded by the National Agency for Research and Development (ANID), Scholarship Program, Doctorado Becas Chile, 2020-21201000. The first author of the article also thanks for the licenses for doctoral studies, the Fund for Teacher and Professional Development of the Universidad Simón Bolívar, Colombia.

References
[1] Logemann J A, Fisher H B, Boshes B, Blonsky E R 1978 Frequency and cooccurrence of vocal tract dysfunctions in the speech of a large sample of Parkinson patients *Journal of Speech and hearing Disorders* **43(1)** 47
[2] Arora S, Baghai-Ravary L, Tsanas A 2019 Developing a large scale population screening tool for the assessment of Parkinson's disease using telephone-quality voice *The Journal of the Acoustical Society of America* **145(5)** 2871
[3] Sakar B E, Isenkul M E, Sakar C O, Sertbas A, Gurgen F, Delil S, Apaydin H, Kursun O 2013 Collection and analysis of a Parkinson speech dataset with multiple types of sound recordings *IEEE Journal of Biomedical and Health Informatics* **17(4)** 828
[4] Wodzinski M, Skalski A, Hemmerling D, Orozco-Arroyave J R, Nöth E 2019 Deep learning approach to Parkinson's disease detection using voice recordings and convolutional neural network dedicated to image classification *41st Annual International Conference of the IEEE Engineering in Medicine and Biology Society (EMBC)* (Berlin: IEEE) p 717
[5] Orozco-Arroyave J R, Arias-Londoño J D, Vargas-Bonilla J F, Gonzalez-Rativa M C, Nöth E 2014 New Spanish speech corpus database for the analysis of people suffering from Parkinson's disease *Proceedings of the Ninth International Conference on Language Resources and Evaluation (LREC'14)* (Reykjavik: European Language Resources Association) p 342
[6] Zahid L, Maqsood M, Durrani M Y, Bakhtyar M, Baber J, Jamal H, Mehmood I, Song O Y 2020 A spectrogram-based deep feature assisted computer-aided diagnostic system for Parkinson's disease *IEEE Access* **8** 35482
[7] Trinh N, Darragh O 2019 Pathological speech classification using a convolutional neural network *Irish Machine Vision and Image Processing (IMVIP 2019)* (Dublin: Technological University Dublin) p 72
[8] Huang G B, Zhu Q Y, Siew C K 2004 Extreme learning machine: a new learning scheme of feedforward neural networks *International Joint Conference on Neural Networks (IEEE Cat. No.04CH37541)* (Budapest: IEEE) p 985
[9] Giuliano M, Perez S N, Maldonado M, Bondar P, Linari D, Adamec D A, Debas M I, Morales C M, León L, Yaco A Y, Birelli J B, Martínez R M, Lacaze M L, Gurlekian J A 2021 Construction of a Parkinson's voice database *International Conference on Industrial Engineering and Operations Management* (Sao Paulo: IEOM Society International) p 940
[10] Greenberg S, Kingsbury B E 1997 The modulation spectrogram: in pursuit of an invariant representation of speech *International Conference on Acoustics, Speech, and Signal Processing* (Munich: IEEE) p 1647
[11] Kingsbury B E, Morgan N, Greenberg S 1998 Robust speech recognition using the modulation spectrogram *Speech Communication* **24** 117
[12] Pietikäinen M 2005 Image analysis with local binary patterns *Scandinavian Conference on Image Analysis (SCIA 2005)* (Berlin: Springer) p 115
[13] Huang G B, Zhu Q Y, Siew C K 2006 Extreme learning machine: theory and applications *Neurocomputing* **70(1-3)** 489
[14] Ding S, Zhao H, Zhang Y, Xu X, Nie R 2015 Extreme learning machine: algorithm, theory and applications *Artificial Intelligence Review* **44(1)** 103
[15] Barata J C A, Hussein M S 2012 The Moore-Penrose pseudoinverse: a tutorial review of the theory *Brazilian Journal of Physics* **42(1-2)** 146
[16] Chamara L, Zhou H, Huang G B, Vong C M 2013 Representational learning with extreme learning machine for big data *IEEE Intelligent Systems* **28(6)** 31
[17] Tang J, Deng C, Huang G B 2015 Extreme learning machine for multilayer perceptron *IEEE Transactions on Neural Networks and Learning Systems* **27(4)** 809
[18] Beck A, Teboulle M 2009 A fast iterative shrinkage-thresholding algorithm for linear inverse problems *SIAM Journal on Imaging Sciences* **2(1)** 183

Prediction of survival of Pinus radiata seedlings subjected to physical-water restriction extreme using learning neural networks

J Vásquez-Coronel[1], A Altamirano-Fernández[1], S Espinoza-Meza[2], and M Rodriguez-Gallardo[1]

[1] Departamento de Matemática, Física y Estadística, Universidad Católica del Maule, Talca, Chile
[2] Departamento de Ciencias Forestales, Universidad Católica del Maule, Talca, Chile

E-mail: jose.vasquez.02@alu.ucm.cl

Abstract. Drought is one of the main environmental factors that limit plant growth. For this reason, it is necessary to apply nursery cultural practices to produce quality seedlings for successful reforestation in drought- prone sites. In this study, the extreme learning machines and multilayer are applied to predict survival in 5-month-old *Pinus radiata* seedlings belonging to 98 families of a genetic improvement program and subjected to a period of water restriction in the nursery. After applying the water restriction, survival was registered in each seedling as a categorical variable (1 = alive seedling, 0 = dead seedling). Additionally, the following morphological attributes of each seedling were also measured: total height, root collar diameter, slenderness index, dry weight of needles, stems and roots, total dry weight, and the root to shoot ratio. The extreme learning machines predicted with a better rate the survival of the "alive" class compared to the "dead" class. On the other hand, the multilayer-extreme learning machines improved the precision of survival concerning the class of "dead" seedlings. According to the results of the model, an overall precision of 74% was obtained. This may be due to the great genetic variability presented by each of the *Pinus radiata* family used in the database. However, this technique allowed predicting the survival of a group of seedlings grown in the nursery, which can be a tool to support the selection process of high quality planting stock.

1. Introduction

In Chile, the main forest companies make great efforts to improve the quality of plantations with *Pinus radiata* (*P. radiata*), a species that sustains the national timber industry and has global scope [1]. However, reduced rainfall and increased temperatures are expected to increase the risk and frequency of drought events in many forested regions of the world [2]. This in turn will cause a high seedling mortality in the summer season after their establishment in the field. In this context, it is necessary to determine the quality of the planting stock at the nursery stage in order to ensure success in the establishment. For this reason it is important to evaluate seedling survival at an early age in the face of a water restriction challenge that simulates field conditions. In this way, plants with a better adaptation to drought-prone sites can be obtained [3].

Espinoza S E, *et al* [4] conducted a nursery study consisting of the application of two irrigation regimes (well-watered versus water restriction), to 5-month-old *P. radiata* seedlings from different geographical and genetic origins. The authors presented an analysis of variance and linear regressions and evaluated phenotypic plasticity in growth and survival at the nursery stage and two years after field establishment. The seedlings had 80% survival in the well- watered treatment, but less than 20% survival in the water restriction treatment. After that, the authors applied a genetic analysis of quantitative traits to estimate genetic parameters associated with growth and biomass allocation in the same *P. radiata* seedlings [5]. However, the use of regression models to predict survival is subjected to a series of restrictions, whereas the use of artificial neural networks (ANN) are among the most widely used machine learning techniques to predict survival and mortality [6].

ANN has been used to predict the survival of *Fagus orientalis* and *Picea abies* using environmental variables, growth, and biomass of adult trees as predictors [6, 7]. In contrast, Şenyurt M, *et al* [8] presented an ANN study to predict the diameter at the breast height from stump diameter and provided alternative modeling for forest management. Guan B T, *et al* [9] developed a model based on an ANN to predict the survival of *Pinus resinosa* trees, using diameter at breast height and annual growth in diameter as predictors. Park Y S [10] presented two different artificial neural networks, both supervised and unsupervised, to assess the risk that *Pinus densioflora* and *Pinus thenbergii* survived an infection caused by the insect *Thecodiplosis japonensis*.

The authors measured eight variables related to the diameter, height, and crown in 251 dead or living trees in a pine forest severely damaged by *T. japonensis.* These eight variables predicted with 87.6% accuracy the state (alive or dead) of the pines. Also, Merkl D, *et al* [11] showed an ANN study to predict individual tree mortality at unequal ages and compare their results with the Logit regression model. The authors found that ANN performs slightly better with respect to the Logit regression model. This corroborates that the application of ANN has great advantages for predicting the survival of a group of plants subjected to biotic or abiotic stresses. However, studies based on the use of ANN to predict survival at an early age in *P. radiata* seedlings or other species of economic interest, are scarce. There is research to predict morphological attributes in nursery-grown *P. radiata* seedlings; however, they are based on other techniques such as artificial vision systems [12].

In this work, the survival of 5-month-old *P. radiata* seedlings subjected to a period of water restriction in the nursery stage is studied. The objective is to predict survival using extreme learning machines (ELM) and multilayer extreme learning machines (ML-ELEM), using attributes of diameter, height and biomass of the seedlings as predictors.

2. Method

This section describes the single-layer and multi-layer ELM architecture learning framework, which was applied to solve the survival prediction problem of *P. radiata* seedlings. The ELM algorithm is a type of feed forward neural network introduced by [13]. This ELM network maps the input data into a hidden layer through random transformations based on randomly generated hidden weights and biases. In the hidden layer, each node summarizes the information of the input layer using a weighted sum as a transfer function [14]; then, the output of this function is sent to an activation function.

The output weights of the ELM network between the hidden layer and the output layer are obtained by applying the Moore-Penrose (MP) inverse matrix [15]. The random generation of the hidden parameters (weights-biases) and the optimization of the network output using the pseudo-inverse are two characteristics that differentiate it from traditional neural networks. Furthermore, the ELM framework is easy to implement computationally, and the algorithm training process is faster than other next-generation classifiers [16]. The following is a description of the ELM model seen from a mathematical context.

Given a training set of N samples, $\{(\mathbf{x}_i, \mathbf{t}_i) \in \mathbb{R}^p \times \mathbb{R}^q\}_{i=1}^{N}$, $\mathbf{t}_i$ is the i-th label associated with the i-th input data $\mathbf{x}_i$. The mathematical model integrated into the ELM framework, which seeks to minimize a distance between the real output of the network and the expected objective, is expressed in Equation (1) [13].

$$f_L(\mathbf{x}_j) = \sum_{i=1}^{L} \boldsymbol{\beta}_i \phi(\mathbf{w}_i \cdot \mathbf{x}_j + b_i) = \mathbf{t}_j, \quad j = 1 \ldots N. \tag{1}$$

It is observed in Equation (1) that the method is generated based on an activation function $\phi(\cdot)$ evaluated at $\mathbf{w}_i \cdot \mathbf{x}_j + b_i$ for a total of L neurons. In this same expression, $\mathbf{w}_i$ and b_i denote the weights and biases of the i-th hidden neuron and $\boldsymbol{\beta}_i$ is the i-th vector of weights at the output of the network. Equation (1) is rewritten in its compact form $\mathbf{H}\boldsymbol{\beta} = \mathbf{T}$, with $\mathbf{H}$, $\boldsymbol{\beta}$, and $\mathbf{T}$ expressed in Equation (2) [16].

$$\mathbf{H} = \begin{bmatrix} \phi(\mathbf{w}_1 \cdot \mathbf{x}_1 + b_1) & \cdots & \phi(\mathbf{w}_L \cdot \mathbf{x}_1 + b_L) \\ \cdots & \cdots & \cdots \\ \phi(\mathbf{w}_1 \cdot \mathbf{x}_N + b_1) & \cdots & \phi(\mathbf{w}_L \cdot \mathbf{x}_N + b_L) \end{bmatrix}, \quad \boldsymbol{\beta} = \begin{bmatrix} \boldsymbol{\beta}_1^T \\ \cdots \\ \boldsymbol{\beta}_L^T \end{bmatrix}, \quad \mathbf{T} = \begin{bmatrix} \mathbf{t}_1^T \\ \cdots \\ \mathbf{t}_N^T \end{bmatrix}. \tag{2}$$

The inputs of the i-th row of the output matrix of the hidden layer $\mathbf{H}$ are the activated signals of the L neurons corresponding to the example i-th $\mathbf{x}_i$. Huang G, *et al* [13] specified the optimal solution of the ELM model through the analytical expression $\boldsymbol{\beta} = \mathbf{H}^{\dagger}\mathbf{T}$, where $\mathbf{H}^{\dagger}$ is the inverse of MP of $\mathbf{H}$ [15]. Other optimization strategies of the optimal $\boldsymbol{\beta}$ value based on the inverse of MP is found in [16]. To conclude a more stable solution and a better generalization state, Huang G, *et al* [14] incorporated the Ridge Regression (RR) theory and the ELM algorithm [17] for the computation of $\boldsymbol{\beta}$ (see Equation (3)).

$$\begin{aligned} &\text{if } N < L: \ \boldsymbol{\beta} = \mathbf{H}^T \left(\tfrac{\mathbf{I}}{\lambda} + \mathbf{H}^T\mathbf{H}\right)^{-1} \mathbf{T}, \\ &\text{if } N \gg L: \ \boldsymbol{\beta} = \left(\tfrac{\mathbf{I}}{\lambda} + \mathbf{H}^T\mathbf{H}\right)^{-1} \mathbf{H}\mathbf{T}, \end{aligned} \tag{3}$$

where $\lambda > 0$ is a regularization term associated with RR and N is the total of samples. Therefore, the resulting output from the ELM network for any training or test proof is as in Equation (4).

$$f_L(\mathbf{x}) = h(\mathbf{x}) \left(\frac{\mathbf{I}}{\lambda} + \mathbf{H}^T\mathbf{H}\right)^{-1} \mathbf{H}\mathbf{T}, \tag{4}$$

where $h(\mathbf{x})$ is the row vector of $\mathbf{H}$ associated with data $\mathbf{x}$. Figure 1 shows a schematic of the proposed ELM framework for the prediction of alive and dead seedlings of *P. radiata* in the nursery. The random parameters in the hidden layer are the weights matrix $\mathbf{W}$ and the bias vector $\mathbf{b}$, while $\boldsymbol{\beta}$ are the network output weights, $\phi(\cdot)$ is the activation function, k indicates the index of the k-th element of vector $\mathbf{x}_i$ and the k-th neuron in the hidden layer, L denotes the number of neurons in the hidden layer, $\mathbf{x}_i$ is the d-dimensional input vector, and $\mathbf{t}_i$ is the m-dimensional target vector corresponding to $\mathbf{x}_i$.

In this paper, we also present an analysis using a multilayer neural network. In particular, an architecture denominated ML-ELM [18]. ML-ELM learns new representations of the input data in the hidden layers through an ELM Autoencoder (ELM-AE) [18], which has the particularity of reconstructing the input signals using random and orthogonal transformations. Just like ELM, the output weights of ELM-AE results from solving the system $\mathbf{H}\boldsymbol{\beta} = \mathbf{X}$ similar to Equation (3), $\mathbf{H}$ is the output matrix of the hidden layer, $\mathbf{X}$ is the output and input data.

The number of ELM-AE is equal to the number of hidden layers in the ML-ELM neural network. We assigned the output weight $\boldsymbol{\beta}^{(i)}$ of the i-th ELM-AE to the i-th hidden layer of

ML-ELM as $(\boldsymbol{\beta}^{(i)})^T$. Now, the output matrix $H^{(i)}$ corresponding to the ML-ELM hidden layer is as in Equation (5).

$$\mathbf{H}^{(i)} = \phi((\boldsymbol{\beta}^{(i)})^T \mathbf{H}^{(i-1)}), \tag{5}$$

where ϕ is an activation function and $\mathbf{H}^{(i-1)}$ is the $(i-1)$-th hidden layer output matrix. Note that for the case $i = 0$, $\mathbf{H}^{(0)} = \mathbf{X}$. To conclude the learning, the unsupervised output representation of the last ML-ELM hidden layer $\mathbf{H}^f$ and the label matrix $\mathbf{T}$ are taken to form the system $\mathbf{H}^f \boldsymbol{\beta} = \mathbf{T}$. Finally, the regularized least squares method is applied to the calculation of the output weights $\boldsymbol{\beta}$ [14], as well as in Equation (3).

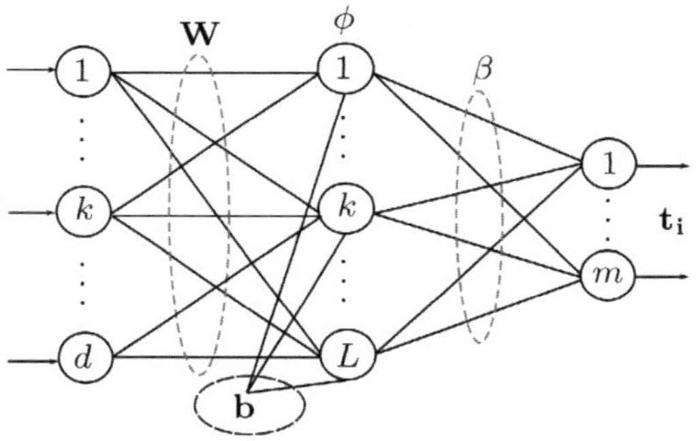

Figure 1. ELM learning architecture.

2.1. Data set

The data set used contains 2,953 samples corresponding to 5-month-old *P. radiata* seedlings belonging to the 98 first, second and third-generation families of a genetic improvement program [19]. The seedlings were cultivated during 5 months in the nursery and then subjected to a water restriction period of 8 weeks. At the end of the water restriction period, the individual survival (SUR) of each seedling was measured as a categorical variable (1 = alive seedling (AS), 0 = dead seedling (DS)). Additionally, the following morphological attributes were also measured in each seedling: total height (H), root collar diameter (RCD), slenderness index (E) as the ratio between H and RCD, dry weight of needles (NDW), stems (SDW) and roots (RDW), total dry weight (TDW), TDW = SDW+ RDW + NDW, and the root to shoot ratio (RSR), RSR = RDW / SDW.

3. Results and discussion

Due to the high seedling mortality after establishment in drought-prone sites, it is necessary to determine the quality of the planting stock for successful forest reforestation. In this context, experimental studies of seedlings submitted to water restriction stress at the early stages of development will give valuable information that facilitate the selection of appropriate seedlings to be established in drought-prone sites. There are some studies that use analysis of variance, linear regression, and quantitative genetics to assess seedling survival in *P. radiata* families submitted to water restriction at the nursery stage [4, 5]. However, statistical analysis is subjected to several limitations (e.g., modeling complex relationships, capturing nonlinear relationships, and dealing with outliers), because they assume a linear-relationship between predictor and response variables.

Consequently, ANN are computational algorithms that do not have these limitations; Bayat M [6] presented an ANN model to predict the survival and mortality of *Fagus orientalis* and *Picea abies* achieving a higher survival rate for plants with diameters in the range of 20 cm - 100 cm. In the same way, Guan B T [9] predicted the survival of *Pinus resinosa* using the diameter at breast height and the annual diameter growth as predictors. Park Y S [10] predicted the survival of *Pinus densioflora* and *Pinus thenbergii*, taking diameter, height and crown of 251 trees as input attributes. Therefore, in this study, we explored the use of ANN, especially the ELM and ML-ELM method, to predict the survival at the nursery stage of *P. radiata* seedlings exposed to a water restriction. The input data is standardized using the Equation (6): for the log-sigmoid activation function in the range $[0, 1]$ and the hyperbolic tangent activation function in the range $[-1, 1]$.

$$X_{\text{norm}} = \frac{X_i - X_{\text{min}}}{X_{\text{max}} - X_{\text{min}}}, \tag{6}$$

where X_i is the value to be standardized, X_{min} and X_{max} is the minimum and maximum value for each input variable, respectively. The network is trained with standardized data to prevent that attributes of larger magnitude have more impact on the learning process [20].

A binary classification problem is solved using a monolayer and multilayer ELM neural network to predict the survival of *P. radiata* seedlings under the abovementioned conditions. To achieve this objective, feature learning methods are applied and evaluated in two scenarios. In the first scenario, the ELM and ML-ELM methods are trained with 70% of the data, which represent 2,066 seedlings. In this training phase, the hyperparameters of the network are adjusted to obtain the best trained model. In the second scenario, known as the testing stage, the results are generalized with the 30% of the remaining samples (886 test examples).

Table 1 describes the ELM and ML-ELM architecture with their activation functions and the input and output variables. Different scenarios are evaluated under a log-sigmoid and hyperbolic tangent activation function defined in Equation (7).

$$f(u) = \frac{1}{1+e^{-u}} \quad \text{y} \quad f(u) = \frac{1-e^{-2u}}{1+e^{-2u}}. \tag{7}$$

Table 1. Input variables used in the ELM and ML-ELM architecture.

Model	Activation functions	Inputs	Outputs
ELM_1	log-sigmoid	RCD, H, E, NDW, SDW, RDW, TDW, RSR	SUR
ELM_2	hyperbolic tangent	RCD, H, E, NDW, SDW, RDW, TDW, RSR	SUR
ELM_3	log-sigmoid	RCD, H, NDW, SDW, RDW, TDW, RSR	SUR
ELM_4	hyperbolic tangent	RCD, H, NDW, SDW, RDW, TDW, RSR	SUR
ELM_5	log-sigmoid	RCD, H, NDW, SDW, RDW	SUR
ELM_6	hyperbolic tangent	RCD, H, NDW, SDW, RDW	SUR
$ML\text{-}ELM_1$	log-sigmoid	RCD, H, E, NDW, SDW, RDW, TDW, RSR	SUR
$ML\text{-}ELM_2$	log-sigmoid	RCD, H, NDW, SDW, RDW, TDW, RSR	SUR
$ML\text{-}ELM_3$	log-sigmoid	RCD, H, NDW, SDW, RDW	SUR

The binary classification results reported in this paper; as the prediction of DSs and ASs, were evaluated from an error matrix defined in Table 2 [21], where true positive (TP) corresponds to number of correctly classified ASs, true negative (TN) is the number of correctly classified DSs, false positive (FP) is the number of DSs incorrectly classified as ASs, and false negative (FN) is the number of ASs incorrectly classified as DSs. In addition, TP + FN is the number of ASs in class one, FP + TN is the number of DSs in class zero, TP + FP is the number of ASs predicted by the network and FN + TN is the number of DSs predicted by the network.

In particular, the performance of the ELM and ML-ELM approach in terms of sensitivity (true positive rate (TPR)), specificity (true negative rate (TNR)), and overall accuracy (OA) is defined in Equation (8).

$$\mathrm{TPR} = \frac{\mathrm{TP}}{\mathrm{TP}+\mathrm{FN}}, \quad \mathrm{TNR} = \frac{\mathrm{TN}}{\mathrm{TN}+\mathrm{FP}} \quad \text{and} \quad \mathrm{OA} = \frac{\mathrm{TP}+\mathrm{TN}}{\mathrm{TP}+\mathrm{TN}+\mathrm{FP}+\mathrm{FN}}, \tag{8}$$

where TRP is the probability that the model identifies an AS, and TNR is the probability of identifying a DS.

Table 2. Binary error matrix associated to the solution of a plant classification problem of *P. radiata* through ELM and ML-ELM.

		Real class		
		AS	DS	Total
Predicted class	AS	TP	FP	TP + FP
	DS	FN	TN	FN + TN
	Total	TP + FN	FP + TN	TP + FN + FP + TN

Table 3 shows the behavior of the statistical indicators associated to an error matrix. It also contains the number of hidden neurons L and the regularization parameter λ required in ELM and ML-ELM training (two hidden layers $L = [L_1, L_2]$). The models reflected in this table correspond to the architectures described in Table 1. From this information, the ELM architectures ($i = 1, \ldots, 6$) have a comparable generalization between them, where the individual accuracy tends to 84% and 63% for ASs and DSs, while the overall accuracy tends to 74%. The error of omission is another way of measuring individual accuracy, which identifies the class to which belong the *P. radiata* assigned. For example for ELM_3, the error of omission for DS is 17.6% (92/523), which indicates that 17.6% is associated to DS. Similarly, the omission error for DS is 37.7% (137/363). Meanwhile, the ML-ELM classifier tends to have an accuracy of 78%, 66% and 73% for TPR, TNR and OA, respectively.

The input attributes in Table 1 allows us to know the dependency relationship between some variables and the impact on the generalization of the ELM and ML-ELM model. The results expressed in Table 3 shows that the class of ASs have a better accuracy rate in the classification process. The accuracy level of ML-ELM compared with ELM improves for the class of DSs. In general, the low rate of accuracy expressed by the ELM and ML-ELM models could be explained by the genetic variation contained by the 98 families of *P. radiata* that participated in the sampling. However, this hypothesis is beyond the scope of this paper and deserves further research.

Table 3. Statistics applied to the evaluation of the ELM and ML-ELM model for predicting survival of *P. radiata* seedlings.

	Error matrix				Accuracy rate			Parameters	
Model	TP	TN	FP	FN	TPR	TNR	OA	L	λ
ELM_1	411	237	142	96	0.8107	0.6253	0.7314	51	$1e+10$
ELM_2	412	240	160	74	0.8478	0.6000	0.7359	1401	$1e-2$
ELM_3	431	226	137	92	0.8241	0.6226	0.7415	801	$1e+4$
ELM_4	428	225	127	106	0.8015	0.6392	0.7370	601	$1e+3$
ELM_5	421	216	143	106	0.7989	0.6017	0.7190	101	$1e+3$
ELM_6	402	241	147	96	0.8072	0.6211	0.7257	1601	$1e-2$
$\mathrm{ML\text{-}ELM}_1$	391	257	130	108	0.7836	0.6641	0.7314	251–71	——
$\mathrm{ML\text{-}ELM}_2$	411	233	118	124	0.7682	0.6638	0.7269	53–95	——
$\mathrm{ML\text{-}ELM}_3$	388	241	135	122	0.7608	0.6410	0.7099	81–71	——

4. Conclusion

In this study we used artificial neural networks to predict the survival of *P. radiata* seedlings subjected to a water restriction challenge in the nursery. The success rate was higher than 74% in the generalization of results. In the individual classification of dead seedling and alive seedling, the extreme learning machine method adopts a better rate of accuracy for the alive seedling class. However, the multilayer extreme learning machines network achieves better performance than the extreme learning machine in classifying the class of dead seedling.The tests show a success rate higher than 64% for dead seedling in the different trained multilayer models. The results show that the neural network can be used to predict seedling survival at t he early stages of development. This is could be an important tool to support the early selection of high quality planting stock. However, in future works the artificial neural networks should be implemented to predict seedling survival according to the morphological attributes and genetic origin of each *P. radiata* seedling.

Acknowledgment

Data used in this study was obtained from the project 'Identification of *Pinus radiata* D. Don families suitable to be established in dryland areas' (Grant number 11121484), funded by the Comisión Nacional de Investigación Científica y Tecnológica from Chile.

References

[1] Espinoza G R, Iribarren O O 2012 *Maderas: Ciencia y Technología* **14(1)** 65
[2] Allen C D, *et al.* 2010 *Forest Ecology and Management* **259(4)** 660
[3] Mead D J 2013 *Sustainable Management of* Pinus Radiata *Plantations* (Rome: Food and Agriculture Organization of the United Nations)
[4] Espinoza S E, Ivković M, Arce M A, Díaz C R, Moya R, Ariza A M 2019 *Scientia Agricola* **77(3)** 1
[5] Espinoza S E, *et al.* 2014 *Tree Genetics & Genomes* **10(4)** 1045
[6] Bayat M, *et al.* 2019 *Computers and Electronics in Agriculture* **164** 1
[7] Hasenauer H, Merkl D, Weingartner M 2001 *Advances in Environmental Research* **5(4)** 405
[8] Şenyurt M, Ercanlı İ, Günlü A, Bolat F, Bulut S 2020 *Bosque* **41(1)** 25
[9] Guan B T, Gertner G 1991 *Forest Science* **37(5)** 1429
[10] Park Y S, Chung Y J 2006 *Forest Ecology and Management* **222(1-3)** 222
[11] Merkl D, Hasenauer H 1998 Using neural networks to predict individual tree mortality *International Conferences on Engineering Application(s) of Neural Network(s)* (Londres: Citeseer) p 197
[12] McGuinness B, Duke M, Au C , Lim S 2021 *Computers and Electronics in Agriculture* **189** 1
[13] Huang G, Zhu Q, Siew C 2006 *Neurocomputing* **70(1)** 489
[14] Huang G, Zhou H, Ding X, Zhang R 2012 *IEEE Transactions on Systems, Man, and Cybernetics, Part B (Cybernetics)* **42(2)** 513
[15] Courrieu P 2008 *Neural Information Processing - Letters and Reviews* **8(2)** 25
[16] Lu S, Wang X, Zhang G, Zhou X 2015 *Intelligent Data Analysis* **19(4)** 743
[17] Hoerl A, Kennard R 1970 *Technometrics* **12(1)** 55
[18] Espinoza S, Magni C, Martinez V, Ivkovic M 2013 *Forest Systems* **22(1)** 3
[19] Kasun L, Zhou H, Huang G, Vong Ch 2013 *IEEE Intelligent Systems* **28(6)** 31
[20] Haykin S 2009 *Neural Networks and Learning Machines* (New Jersey: Pearson Education Inc.)
[21] Visa S, Ramsay B, Ralescu A, Van Der Knaap E 2011 *The Modern Artificial Intelligence and Cognitive Science Conference* **710** 120

Impulsive simulation model for the dynamics of allergen immunotherapy

J G Vergaño-Salazar[1], F Córdova-Lepe[1,2], L Pastenes[1,2], L Cuesta-Herrera[1], and R Lozada-Yavina[1]

[1]Doctorado en Modelamiento Matemático Aplicado, Universidad Católica del Maule, Talca, Chile

[2]Facultad de Ciencias Básicas, Universidad Católica del Maule, Talca, Chile

E-mail: jgvs190@gmail.com, juan.vergano@alu.ucm.cl

Abstract. This study aims to analyze the effects of allergen immunotherapy, used to treat allergic symptoms such as pollen allergy. Mathematical models are used as a methodological approach to simulate from a system of impulsive differential equations the dynamics of the model. Immunotherapy is based of supplying small amounts of pollen to the patient, which leads to minimizing severe allergic symptoms when patients are subsequently exposed to higher amounts of pollen in the environment. Lymphocyte concentrations are considered state variables, allowing the behavior and efficacy of allergen immunotherapy to be identified. The manuscript proposes a method that allows to model mixed systems. Phenomena that present continuous times in some instants and discrete times in others, these are phenomena that are frequently found in the field of physics. Allergen immunotherapy is most effective when a treatment is created with pollen dose increments in a linear form.

1. Introduction

Allergy is a disorder caused by an immediate hypersensitivity reaction, often referred to the type of antigen (allergen) that triggers the disease. Exposure can be by ingestion, inhalation or direct contact. All of these disorders are the result of immunoglobulin E (IgE) production stimulated by T helper (Th) lymphocytes, followed by allergen-dependent mast cell activation and IgE [1, 2].

T lymphocytes play an important role in immune responses to allergenic substances [3]. In the case of allergy, it follows that there are allergen-specific T lymphocyte populations in both allergic and healthy individuals [4]. An allergic reaction begins with stimulation of the immune system through sensitization during which allergen-specific T lymphocytes develop. The dynamics are described as follows: (1) antigen presenting cells (APC) receive fractions of antigens of the major histocompatibility complex (MHC I) and process them into peptides. (2) The APCs present the peptides to naive T lymphocytes (Th_0), which differentiate into lymphocytes (Th_2) in the presence of interleukin 4 ($IL-4$). (3) Th_2 lymphocytes activate B lymphocytes (LB) and these once produce (IgE) which binds to mast cell receptors. (4) Mast cells release histamine which induce inflammation and allergic symptoms [1, 5]. Once this dynamic unfolds, the regulatory T lymphocytes ($Treg$) suppress the allergic response [1,6,7].

The mechanisms of immune suppression by allergen immunotherapy remain poorly understood [8]. Studies suggest that $Treg$ cell immunosuppressive cytokines such as $IL-10$

and transforming growth factor $\beta(TGF-\beta)$ as secreted cytokines and various surface molecules play multiple roles in suppressing immune responses and directly inhibiting the development of inflammatory responses [2,9–11]. In addition, $Treg$ lymphocytes suppress immune response by preventing APCs from activating Th_0 lymphocytes [7,12]. Finally, it should be taken into account that the number of $Treg$ lymphocytes can be modified by the presence of infections [13].

Allergen immunotherapy can restore tolerance to allergens [7,14]; many of the immunological measurements support a reduced $CD4+Th_2$ lymphocyte response [7, 11, 14]. Type I hypersensitivity is inappropriately induced by foreign substances of protein origin such as pollen. The immune response in the lymph nodes begins when pollen enters the organism through the mucosal surfaces. This response becomes dominated by a subset of lymphocytes Th_2 [1,15].

The predominance of Th_2 leads to the production of IgE, which makes it a marker indicator of the allergic response. Immunotherapy presumably introduces the allergen into the non-mucosal lymph nodes and can lead to a Th_1 dominated response leading to the production of immunoglobulin G (IgG) [15]. Anaphylaxis, mediators derived from mast cells or basophils can restrict the airways to the point of suffocation and cause cardiovascular collapse [1]

Mathematical models to describe the dynamics allergy dynamics have used different tools and approaches. The manuscript proposes a method that allows to model mixed systems. Phenomena that present continuous times in some moments and discrete times in others. Therefore, a tool is presented that allows modeling of quivalent systems present in the field of physics.

Taking into account that the dynamic system is at the microscopic level, computational tools are used in this work to represent the mathematical model proposed. Therefore, the objective of this work is to build a model that describes the immunotherapy treatment with allergens and the response of the immune system to it. Through a system of impulsive differential equations and making use of scientific computing, it is possible to analyze and interpret different simulation scenarios of the proposed model.

2. Mathematical modeling

This study presents a mathematical model that describes allergen immunotherapy treatment in people with pollen allergies. The methodology based on a hypothetical-deductive structure typical of mathematical modeling [16,17] where the selection of variables, parameters and the conjecture of the relationships in the interaction dynamics between lymphocytes Th_0, $Treg$ and Th_2. This allows the construction, simulation and interpretation of a model based on impulsive differential equations that describe said treatment.

To show the effects of allergen immunotherapy and its results when the patient is exposed to high concentrations of pollen in the environment, the dynamics are described on two time scales, continuous and discrete, typical of the formalism of impulsive differential equations. [18,19]. The first describes the response of the immune system in the absence of the allergen either by therapy or by exposure $t \neq \{t_k, t_m\}$ phenomenon described by cell interaction, where the naive lymphocytes or Th_0 change their concentration due to their differentiation processes, both for $Treg$ lymphocytes and Th_2 lymphocytes, with rates α_2, α_3 respectively. The half-life death rate is μ_1. On the other hand, the $Treg$ lymphocytes grow at a rate α_2 as a result of the differentiation process of the Th_0 lymphocytes and decrease with a mortality rate μ_2 due to their half-life. In the case of Th_2 lymphocytes, these grow at a rate α_3 as a result of the differentiation process of Th_0 lymphocytes, but are regulated by $Treg$ lymphocytes at a rate m.

The number of $Treg$ cells resulting from the phenomenon of lymphopoiesis (creation of $Treg$ cells by the thymus) is n and the mortality rate of $Treg$ lymphocytes product of the half-life is μ_3. This process is carried out from an initial time $t_0 = 0$ to an end time t_f and is only disturbed by the impulses resulting from the doses of immunotherapy with allergens or by exposure to concentrations of allergens in the environment.

The second (discrete time), presented in two phases, A: phase of immunotherapy with allergens or desensitization, where the patient is inoculated with low doses of allergen for a period $[0, T]$. This period allows the desensitization of the immune system. This process is carried out in the form of a pulse at times $t = t_k$, following an increasing sequence of times $\{t_k\}$ such that $t_{k+1} = t_k + \tau$ for $\tau \geq 0$ and $\tau_0 = 0$, where the immune system responds to this stimulus.

In this case, the Th_0 lymphocytes grow as a result of the concentration of allergens supplied in dose "a" and the number of Th_0 lymphocytes per pollen unit α. B: phase of exposure to allergens in the environment, which is described by a daily concentration $P(t)$ in the form of a pulse at times $t = t_m$, following an increasing succession of times t_m. In this way, the interaction model between the different state variables is schematically represented in Figure 1. Where the lines represent the cellular interaction and the underlying dynamics in the mathematical model.

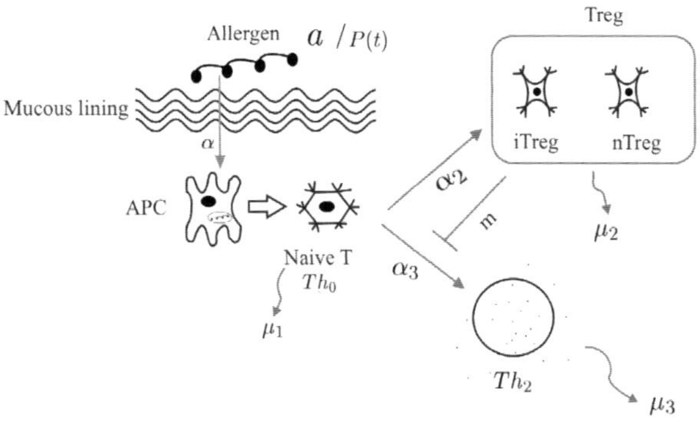

Figure 1. Conceptual model for allergen immunotherapy treatment. Adapted from [16].

The variables and parameters included in the construction of the model that describes the allergen immunotherapy treatment are shown in Table 2.

Table 1. Variables and parameters of the interaction dynamics between pollen allergen and immune system.

Description	Variable	Unit
Number of naive T lymphocytes	Th_0	CU
Number of regulatory T lymphocytes	$Treg$	CU
Number of type-2 T helper lymphocytes	Th_2	CU
Mortality rate of Th_0 lymphocytes (half life)	μ_1	$1/day$
Rate of differentiation of Th_0 lymphocytes into $Treg$ lymphocytes	α_2	1
Rate of differentiation of Th_0 lymphocytes into Th_2 lymphocytes	α_3	1
Mortality rate of $Treg$ lymphocytes (half life)	μ_2	$1/day$
Regulatory rate of $Treg$ lymphocytes over Th_2	m	1/CU
Number of $Treg$ lymphocytes produced by the Thymus	n	CU
Mortality rate of Th_2 lymphocytes (half life)	μ	$1/day$
Number of differentiated Th_0 lymphocytes per pollen unit (PU)	α	1
Pollen doses	a	PU/day
Pollen concentration curve	$P(t)$	PU/day

After the hypotheses raised, a mathematical model based on a system of impulsive differential equations is proposed, thus describing the immunotherapy treatment with allergens as shown by Equation (1).

$$\begin{cases} \left.\begin{aligned} Th_0'(t) &= -\mu_1 Th_0(t) - \alpha_2 Th_0(t) - \alpha_3 Th_0(t) \\ Treg'(t) &= \alpha_2 Th_0(t) - \mu_2 Treg(t) \\ Th_2'(t) &= \frac{\alpha_3 Th_0(t)}{1 + m(Treg(t) + n)} - \mu_3 Th_2(t) \end{aligned}\right\} \text{ if } t \neq \{t_k, t_m\}, t \in (0, t_f), \\ \left.\begin{aligned} Th_0(t^+) &= Th_0(t) + \alpha a \\ Treg(t^+) &= Treg(t) \\ Th_2(t^+) &= Th_2(t) \end{aligned}\right\} \text{ if } t = t_k, t \in (0, T], \\ \left.\begin{aligned} Th_0(t^+) &= Th_0(t) + \alpha P(t) \\ Treg(t^+) &= Treg(t) \\ Th_2(t^+) &= Th_2(t) \end{aligned}\right\} \text{ if } t = t_m, t \in (T, t_f). \end{cases} \tag{1}$$

3. Results and discussion

This work considers two simulation scenarios related to the way of delivering the immunotherapy treatment. In each scenario there are two phases, the treatment phase 180 days (6 months) and the phase of exposure to the allergen in the environment. the first scenario, with a fixed dose throughout the treatment period and the second, with a linear growth, starting with a minimum dose to gradually reach high concentrations of the allergen, which do not generate allergic reactions, the latter is called maintenance dose [20]. The allergen concentration in the exposure periods corresponds to a pollen curve that approximates the pollen concentration of *Platanus orientalis* during the months of August to September as shown in Figure 2. Approximate curve for the first annual record of allergenic pollen in Talca, Chile [21].

For the different simulation scenarios the parameters $\mu_1 = 0.001$, $\alpha_2 = 0.5$, $\alpha_3 = 0.5$, $\mu_2 = 0.001$, $m = 0.045$, $n = 200$, $\mu_3 = 0.01$, $\alpha = 10$ are kept constant and are taken from secondary sources [5]. With these conditions, it is found that the adequate treatment to avoid the allergic response must have a total dose of 180 PU.

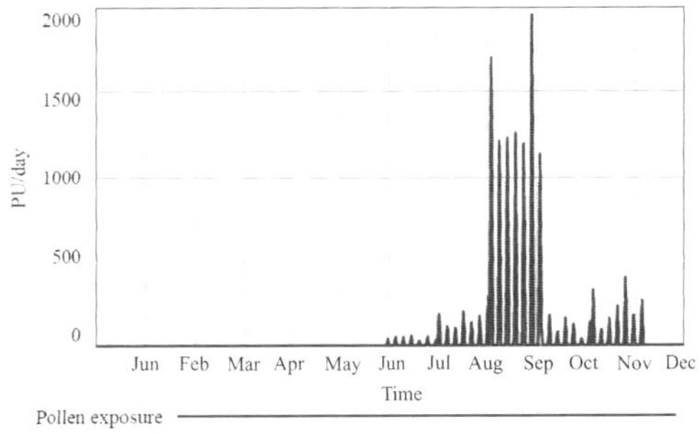

Figure 2. Pollen curve: pollen concentration of *Platanus orientalis*.

3.1. Treatment with a constant dose of pollen

Figure 3, Figure 4, Figure 5 and Figure 6 are the result of inoculating constant doses throughout the treatment period. The allergic response threshold (blue line) with a concentration of Th_2 lymphocytes equal to 50 for all cases, it should be noted that exceeding 80% of the threshold of allergic response (red line), with a concentration of Th_2 lymphocytes equal to 40 for all cases, will indicate that the organism is in danger, therefore, it is sought that the concentration of lymphocytes Th_2 with immunotherapy treatment (green line) is below this threshold. It is evident that the concentration of lymphocytes Th_2 without treatment (gray line) exceeds these thresholds.

Journal of Physics: Conference Series **2153** (2022) 012016 doi:10.1088/1742-6596/2153/1/012016

The time in months is shown on the horizontal axis corresponds to 365 days from January 1, for which there are doses corresponding to the treatment until day 180, this includes the months of January to June. The vertical axis shows the number of cell concentration Th_2. A similar trend is evidenced in the body at approximately 190 days regardless of whether or not it is with immunotherapy treatment.

Figure 3, Figure 5 and Figure 6 show an amount of 36 doses of 5 PU/day, 9 doses of 20 PU/day and 6 doses of 30 PU/day, respectively, which corresponds to a treatment with a total of 180 PU. In general, a similar behavior is shown in all the simulation scenarios and it is evident that all the concentrations of Th_2 are above 80% of the threshold of the allergic response. This indicates that for these parameter conditions the appropriate doses have been found. The organism is prone to higher concentrations of Th_2, without considering that it is in danger since it does not reach or exceed the threshold for allergic response (blue line), reaching a maximum concentration of 42 Th_2 UP.

Figure 4 shows an amount of 15 doses of 12 PU/day, this being an adequate concentration to not reach or exceed the Th_2 lymphocyte threshold corresponding to 80%. This is considering that is the conditions that they are desired to achieve so that the organism is out of danger before an allergic reaction.

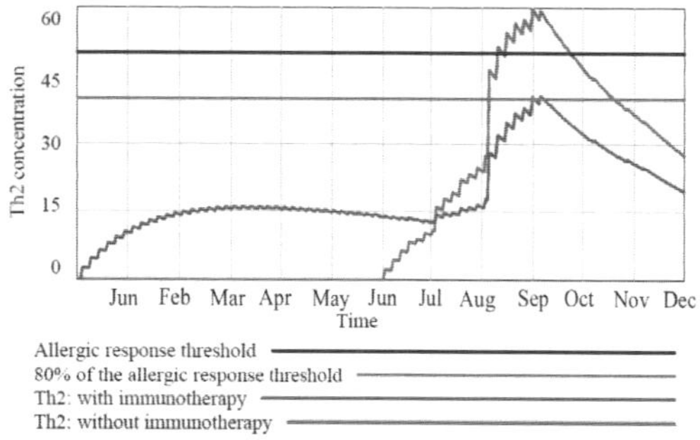

Figure 3. 36 doses of 5 PU/day.

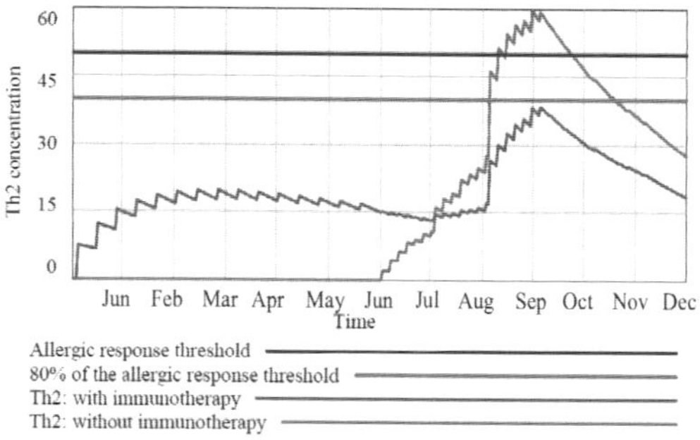

Figure 4. 15 doses of 12 PU/day.

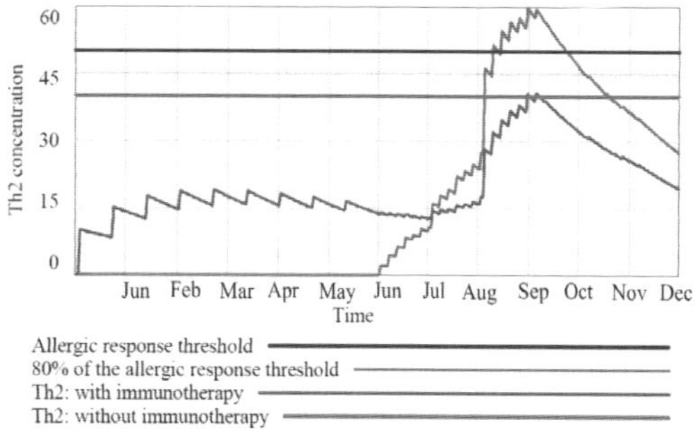

Figure 5. 9 doses of 20 PU/day.

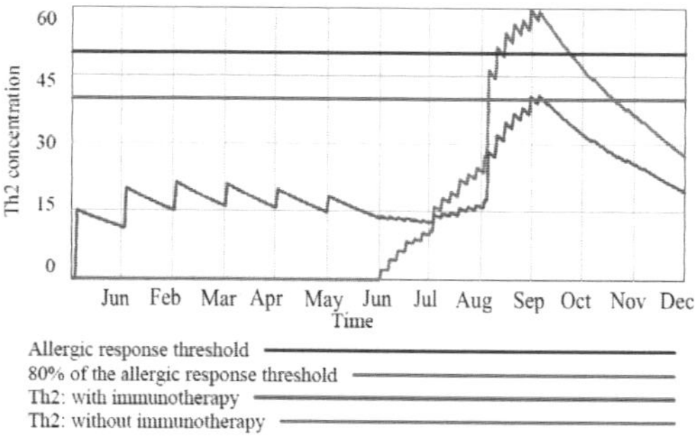

Figure 6. 6 doses of 30 PU/day.

3.2. Treatment with linear pollen dose increments

This simulation scenario presents the result of allergen immunotherapy where the treatment dose is described by a linear growth between doses. This begins with a minimum dose to gradually reach the maintenance dose as shown in Figure 7. Figure 8 shows the results of using the treatment where the threshold of allergic response is 50 (blue line) and 80% of this (red line) represents the threshold of allergy danger. The concentrations of Th_2 without treatment (gray line) and with treatment (green line) show that immunotherapy treatment with allergens is more effective than that described by a constant dose.

Figure 3 and Figure 8 show a treatment with 36 doses. The first is the case of constant dose and the second is the case of linearly increasing dose. the latter case, it is evidenced that immunotherapy treatment with allergens presents a better response in the process of desensitization of the immune system, that is, immunotherapy is more effective when a treatment is carried out with increases in pollen doses in a linear manner. This allows the patient's response to be better in the presence of high concentrations of allergen in the environment.

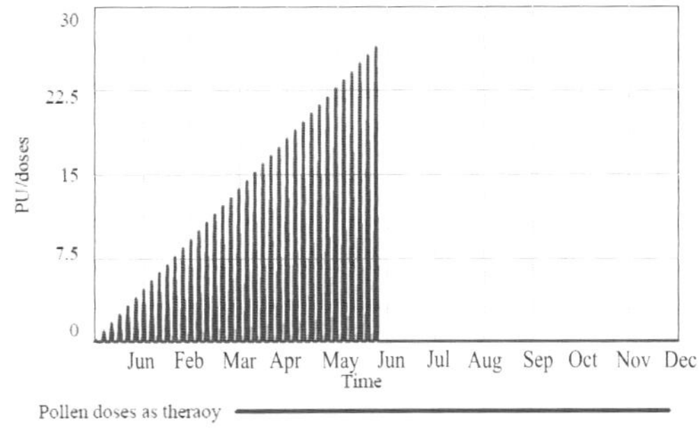

Figure 7. Treatment with linear pollen dose increments.

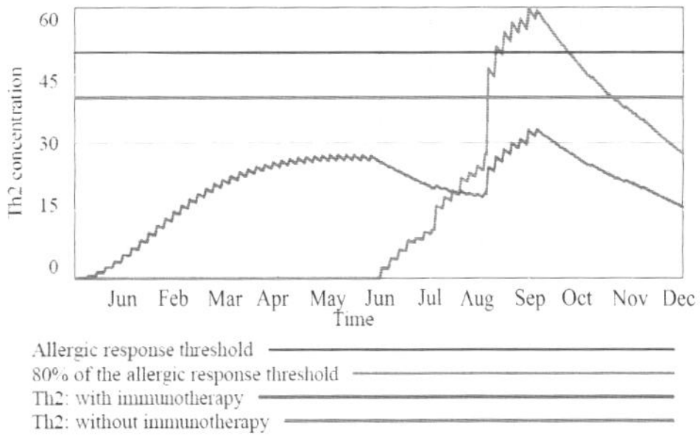

Figure 8. Linear treatment effect.

4. Conclusions

The study shows how mathematical modeling allows to simulate complex systems with highly non-linear relationships at the microscopic level, where the particles obey a simplified dynamics. The manuscript proposes a method that allows to model mixed systems. Phenomena that present continuous times in some instants and discrete times in others, contributing to the modeling of equivalent systems present in physical phenomena.

Furthermore, the study shows the efficacy of allergen immunotherapy treatment in patients with hypersensitivity to pollen allergens. After identifying the parameters α, α_2 and α_3 in an individual, the model made it possible to find the appropriate dose so that the concentration of lymphocytes Th_2 remains below 80% of the threshold of allergic response. Furthermore, it is concluded that allergen immunotherapy treatment may be more effective if the delivered dose follows a linear growth.

Acknowledgements

Juan Gabriel Vergaño Salazar, Ledyz Cuesta Herrera and Rafael Lozada Yavina thanks the "UCM Doctoral Scholarship", awarded by the Vice-Rectory for Research and Postgraduate of the Universidad Católica Del Maule, Talca, Chile.

References

[1] Abbas A K, Lichtman A H, Pillai S 2014 *Cellular and Molecular Immunology* (Philadelphia: Elsevier Health Sciences)
[2] Cappella A, Durham S 2012 Allergen immunotherapy for allergic respiratory diseases *Human Vaccines & Immunotherapeutics* **8(10)** 1499
[3] Zhu J, Paul W E 2010 Heterogeneity and plasticity of T helper cells *Cell Research* **20(1)** 4
[4] Romagnani S 1997 The th1/th2 paradigm *Immunology Today* **18(6)** 263
[5] Hara A, Iwasa Y 2017 When is allergen immunotherapy effective? *Journal of Theoretical Biology* **425** 23
[6] Sakaguchi S, Yamaguchi T, Nomura T, Ono M 2008 Regulatory T cells and immune tolerance *Cell* **133(5)** 775
[7] Heldner A, *et al.* 2021 Immunological effects of adjuvanted low-dose allergoid allergen-specific immunotherapy in experimental murine house dust mite allergy *Allergy* **00** 1
[8] Collison L W, Workman C J 2008 How regulatory T cells work *Nat Rev Immunol* **8(7)** 523

[9] Taylor A, Verhagen J, Blaser K, Akdis M, Akdis C A 2006 Mechanisms of immune suppression by interleukin-10 and transforming growth factor-β: the role of T regulatory cells *Immunology* **117(4)** 433
[10] Akdis M, *et al.* 2004 Immune responses in healthy and allergic individuals are characterized by a fine balance between allergen-specific T regulatory 1 and T helper 2 cells *Journal of Experimental Medicine* **199(11)** 1567
[11] Wambre E 2015 Effect of allergen-specific immunotherapy on CD4+ T cells *Current Opinion in Allergy and Clinical Immunology* **15(6)** 581
[12] Wing K, *et al.* 2008 CTLA-4 control over Foxp3+ regulatory T cell function *Science* **322(5899)** 271
[13] Jaramillo F J, Gómez L M, Anaya J M 2011 Células T reguladoras, infección y autoinmunidad: implicaciones en terapéutica *Infectio* **10(3)** 178
[14] Greenberger P A 2002 Immunotherapy update: mechanisms of action *Allergy and Asthma Proceedings* (Key Biscayne: OceanSide Publications) p 373
[15] Fishman M A, Segel L A 1996 Modeling immunotherapy for allergy *Bulletin of Mathematical Biology* **58(6)** 1099
[16] Vergaño-Salazar J, Pastenes L, Córdova-Lepe F, Mardones-Precht P 2020 Impulsive simulation model for the analysis of allergy dynamics *Journal of Physics: Conference Series* **1702(1)** 012009:1
[17] Salazar-Viedma M, Vergaño-Salazar J G, Pastenes L, D'Afonseca V 2021 Simulation model for hashimoto autoimmune thyroiditis disease *Endocrinology* **162(12)** bqab190
[18] Córdova-Lepe F, Pinto M, González-Olivares E 2012 A new class of differential equations with impulses at instants dependent on preceding pulses. Applications to management of renewable resources *Nonlinear Analysis: Real World Applications* **13(5)** 2313
[19] Córdova-Lepe F, Robledo G, Cabrera-Villegas J 2015 Population growth modeling with boom and bust patterns: the impulsive differential equation formalism *Journal of Biological Systems* **23(supp01)** S135
[20] Sánchez J, *et al.* 2016 Allergen immunotherapy: mechanisms of action, and therapeutic and socioeconomic impact consensus of the Asociación Colombiana de Alergia, Asma e Inmunología *Biomédica* **36(3)** 463
[21] Mardones P, *et al.* 2013 First annual register of allergenic pollen in Talca, Chile *Allergologia et Immunopathologia* **41(4)** 233

Journal of Physics: Conference Series **2153** (2022) 012017 doi:10.1088/1742-6596/2153/1/012017

Mathematical model and experimental validation for the prediction of dissolved oxygen saturation in aquaculture ponds

K M Valdez-Prudencio[1], S Arceo-Diaz[1], J A Bricio-Barrios[2], and E E Bricio-Barrios[1]

[1] Tecnológico Nacional de México, Colima, México
[2] Universidad de Colima, Colima, México

E-mail: santiago.arceo@colima.tecnm.mx, elena.bricio@colima.tecnm.mx

Abstract. Fluid mechanics is one of the oldest areas of physics with the greatest number of applications in everyday life. This area became more versatile when mass and energy transport equations were incorporated. Together, these equations allow describing a wide variety of case studies with great precision. Among cases available in the open literature, aquaculture is one of the most important due to the growing need for food sources for human consumption, the nutritional value of many fish varieties, the low cost of the maintenance of fish farms. Dissolved oxygen is one of the most relevant parameters ensuring water quality in fish farming. Many fish farms use permanent artificial aeration systems for maintaining the oxygen level within the recommended range used, causing an increasing energy consumption. Therefore, this work proposes an equation and validation, based on equations hydraulics and transport phenomenon, capable of determining the maximum concentration of dissolved oxygen in the body from the operating parameters of the tilapia rearing pond.

1. Introduction

Fluids mechanics describes the movement of fluids in many natural phenomena and the forces that produce this motion. This branch of classical physics is of great importance for engineering since it allows describing the movement of fluid inside pipes traditionally used in transport and pumping systems [1].

By having an approximate mathematical description of the behavior of the fluid, it is possible to propose modifications in the interior of the pipe to regulate flow speed and acceleration, the pressure and temperature of the fluid, and the output volume [2]. Venturi tubes are one example of this. The geometry of Venturi tubes corresponds to two conic sections joined by a thin cylindrical tube, through which the speed of a passing fluid increases (causing a drop in the local pressure). Also, if an aperture is made in the thin cylindrical joint, the pressure difference will produce a suction that can be used to inject oxygen into the passing fluid [3].

The modified Venturi tube allows the fluid to be oxygenated in such a way that it has been used in the medical [4], environmental [5], mining [6] and other areas. In the case of aquaculture, Venturi tubes are widely used as components in many aeration systems for tilapia rearing ponds [7]. The concentration of dissolved oxygen (DO) is a determinant factor on the stress level of specimens during the rearing process [8]. For example; if the concentration of DO is either less than 3 mg/lt or greater than 10 mg/lt will make the specimens susceptible to stress, poor appetite, and growth [9] and, sometimes, severe skin diseases [10].

Although aeration systems based on Venturi tubes allow a reduction in the energy cost related to oxygenation, the maximum concentration of DO in the body of water also depends on the weather conditions, the growth stage of the specimen, and the water quality parameters [11]. This dependence on multiple factors makes it necessary to identify a strategy to calculate the maximum concentration of DO in the thank's water [12].

This work proposes an equation to calculate the maximum concentration of DO in the body of water, deriving it from the theoretical description of the hydraulics of the Venturi tube and the mass transfer rate of oxygen into the body of water. We validated the model's prediction by comparing the theoretical oxygen level to a workbench available in the open literature and the readings on the oxygen concentration obtained from an artificial water compression system, coupled to a Venturi tube.

2. Methodology

To obtain the equation that describes the DO saturation rate in the body of water; this work was carried out in the steps shown below.

2.1. Equation for the concentration of dissolved oxygen

The Equation (1) of the DO rate, with respect to time, is described through a first-order differential mass transfer equation from American Society of Civil Engineers (ACSE) in [13]:

$$\frac{dC}{dt} = K_{La}(C_{sat} - C_0), \quad (1)$$

where C_0, C and C_{sat} correspond to the different concentrations for the DO: the initial DO concentration, the concentration at the aquatic environment and the saturation concentration; respectively (all measured in mg/lt) and k_{La} is the mass transfer rate per unit time (measured in h^{-1}). By solving Equation (1) for k_{La}, we obtain the Equation (2).

$$K_{La} = ln\left[\frac{C_{sat} - C_0}{C_{sat} - C}\right] * \frac{1}{t}. \quad (2)$$

It is important to mention that the transfer rate constant (k_{La}) is given at 20 $°C$ and, in case that the experiment is carried out at a different temperature, the following correction factor is required, Equation (3).

$$K_{La20} = K_{La}(1.024)^{20-T}, \quad (3)$$

T is represents water temperature measured in Celsius degrees. Once the DO saturation of the water has been reached, the efficiency of DO transfer can be estimated by the Equation (4).

$$E = \frac{R}{S} * 100\%, \quad (4)$$

R is the mass rate of oxygen entrained into water body and S is the mass rate of oxygen supplied by the air compressor (both measured in kg/h); these two parameters are defined by the Equation (5) and Equation (6).

$$R = (K_{La}(1.024)^{20-T}) * 8.84 * hd^{0.11} * V * (10^{-3}), \quad (5)$$

$$S = 0.3Q_g, \quad (6)$$

where hd represents the depth of immersion of the diffuser (measured in m), V is the volume contained in the body of water (measured in m^3 and Q_g is the flow of air generated by the compressor (measured in m^3/hr).

2.2. Air flow equation
Traditionally, the magnitude of Q_g is obtained from the compressor; however, this electromechanical system is not capable of oxygenating the fluid inside the tube. Therefore, it is required to incorporate a Venturi tube that allows suction of air from outside trough a conduit perpendicular to the tube; the fluid outlet velocity can be calculated by the Equation (7).

$$v_1^2 * A_1^2 = v_2^2 * A_2^2, \tag{7}$$

and, in turn, this allows to calculate the flow generated by the Venturi tube, Equation (8).

$$Q_2 = A_2 * \sqrt{\left(\frac{2*(P_2 - P_1}{\rho}\right) * \left(1 - \left(\frac{A_1}{A_2}\right)^2\right)}, \tag{8}$$

Q_2 is the fluid's outlet flow (measured in m^3/h). This result is analogous for Q_g. While v_i and P_i correspond to the velocity and pressure at area of entry ($i = 1$) and exit ($i = 2$) of the Venturi tube in (measured in m/s and Pa, respectively) and ρ is the fluid's density (measured in kg/m^3).

3. Results and discussions
The proposed mathematical model has the objective to approximate the magnitude of the saturation oxygen concentration, this work was developed in the three stages shown below.

3.1. Outlier data evaluation and workbench
The data used for the validation of the proposed model was extracted from the work by Cheng X, *et al.* [14], who performed 54 experiments varying the operating characteristics in a tank with different volumes of water body, fluid outlet flows, depth of immersion of the diffuser, bubble diameter, saturation DO concentration, speed of transfer, superficial speed of transfer of the bubbles towards the body of water and in response the efficiency of the process. The data was subjected to a search for atypical data, through by the box-and-whisker plot method, and it was determined that the test marked as "12" exceeded the upper whisker of the distribution.

3.2. Design, construction and validation of the process efficiency equation
Once the statitical reliability was confirmed, a third-order factorial regresion equation from the three main variables (called E_{03}) that affect the efficiency of the process (hd, Q_g and Vol) was developed in Equation (9). The Equation (9), it becomes obvious that some factors, that could be tough as relevant, do not have a significant impact on the final efficiency (as their corresponding coefficients on are equal to zero).

$$\begin{aligned} E_{03} &= 16.398 + 0 * hd - 8.8679 * Q_g + 0 * Vol - 6.6383 * hd * Q_g - 9.8508 * hd * Vol \\ &+ 0.9111 * Q_g * Vol + 2.7666 * hd * Q_g * Vol. \end{aligned} \tag{9}$$

Given that sometimes it is not possible to identify the magnitude of the compressor flow rate or there is no equipment to measure the inlet and outlet pressure to estimate this variable indirectly. Therefore, the construction of a second-order factorial regression equation (called E_{O2}) that depends only on the depth of submersion of the diffuser and the volume of the water body was proposed in Equation (10).

$$E_{O2} = 1.9004 + 0 * hd + 5.199 * Vol - 1.2935 * hd * Vol. \tag{10}$$

Figure 1 shows a contour diagram that represents the efficiency of the DO concentration, calculated by Equation (10), as a response variable on the diffuser's depth (horizontal axis) and the volume of water (vertical axis). Each colored region corresponds to a specific range for the efficiency of DO within 5% and 7.5% and results from selecting different combinations for the diffuser's depth and the water volume of the container. For example, for a specific volume of water, if the diffuser is located near the surface, the efficiency of DO injected into the water will be low (represented by the dark-blue region going in diagonal on the diagram's lower edge). If the body of water has a high volume, the maximum efficiency for oxygen diffusion will be reached with a depth between 0.3 m and 0.6 m.

On the other hand, for a volume between 0.7 m^3 and 0.8 m^3, and a depth of 0.9 m, an efficiency of approximately 7.3% will be achieved. From the experimental data of Cheng X, *et al.* [14] and from the factorial regression Equation (9) and Equation (10), it was identified that the average percentage error of these equations was 4.52% and 4.63%, corresponding to 0.452 mg/l and 0.463 mg/l, respectively.

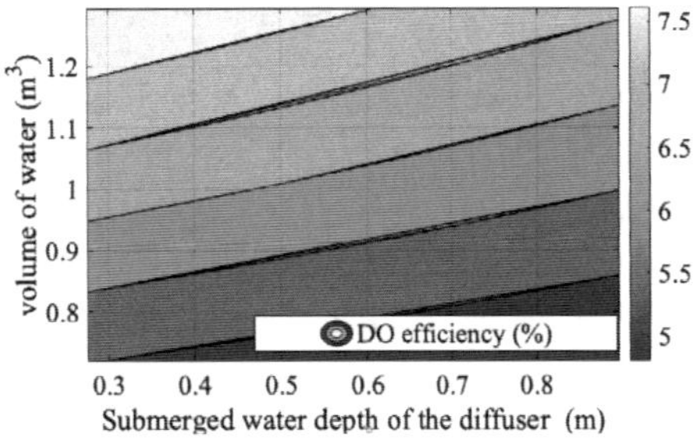

Figure 1. Contour diagram that associates physical parameters and the efficiency of the oxygenation process of the water medium.

3.3. Construction of the saturation dissolved oxygen concentration model

Once the equation allowing the calculation of DO efficiency from the characteristics of the aeration system from Equation (10) has been obtained, the next stage was to correlate the variables that describe the maximum concentration of DO in the body of water. Equation (11) was obtained by equating the factorial regression equations for DO efficiency with the mass transfer rate of Equation (2), Equation (5) and Equation (6) and, by solving for the saturation DO concentration (C_{sat}), we obtain Equation (11).

$$C_{sat} = \frac{exp\left(\left[\frac{[E_i]t*0.3*Q_g}{1.024^{(20-T)}*8.84*hd^{0.11}*V*10^{-3}}\right]\right) * C + C_0}{exp\left(\left[\frac{[E_i]t*0.3*Q_g}{1.024^{(20-T)}*8.84*hd^{0.11}*V*10^{-3}}\right]\right) * C - 1}, \tag{11}$$

where E_i corresponds to the second or third order factorial regression equation. To determine if the mathematical model can be considered as viable. We validated Equation (11) through the comparison between the predictions of the regression Equation (9) and Equation (10) with experimental datas from Cheng X, *et a.l* [14]. Also, we evaluated the average percentual error, giving 9.32%, for Equation (9), and 9.43%, for Equation (10). These values correspond to 0.932 mg/l and 0.943 mg/l of DO in the body of water, respectively.

Finally, the proposed mathematical model was validated by applying a pumping system, coupled to a Venturi tube designed and built by galvanized steel tubes and reductions, a commercial pump with 380 $Watts$ of power, and a plastic tube that will perform the function of the diffuser, put at a depth of 0.1 m. Meanwhile, the DO and temperature measurement and monitoring module used EZO and Atlas Scientific ENV-SDS brand probe integrated circuits, respectively.

Regarding the monitoring system, an Arduino 1 microprocessor and free access software were used to create a timestamp and data storage obtained from the measurement module. Once the compression system and sensors were calibrated, an oxygenation test was performed on a 3-liter body of water. The aeration system and aeration paths are shown in Figure 2.

The Figure 2(a) shows the flow diagram used to aeration system that performed the oxygenation of the water body. The values for the DO concentration and temperature were recorded at one-minute intervals. Once the experiment concluded, the stored data were used to plot the graph shown in the Figure 2(b), to determine the performance of the proposed mathematical model with the second and third-order Efficiency equations. It was observed that the concentration of experimental DO and those obtained by the mathematical model have an error of 1 mg/lt, this value remained almost constant through the experiment. This suggests that the equation proposed for the saturation concentration can be considered as an approximation that must be reinforced through a correction factor or by incorporating more variables into the efficiency equations constructed through factorial regression.

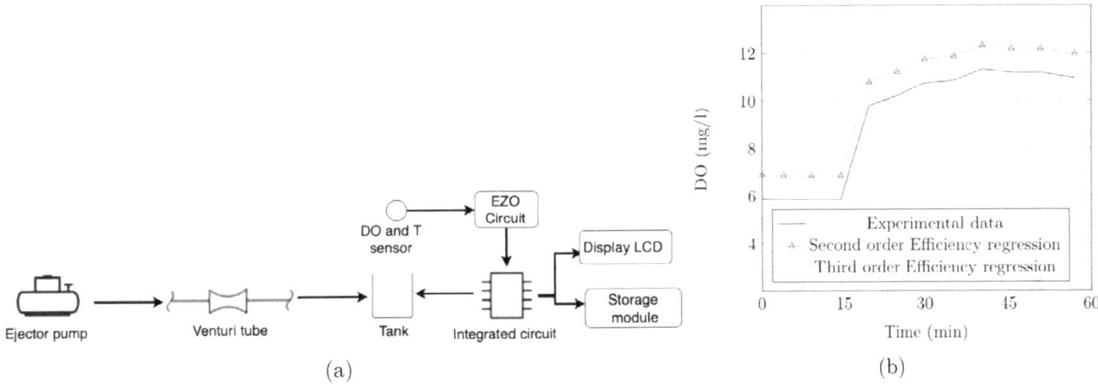

Figure 2. (a) Flow diagram of the Venturi tube oxygenation system, measurement and storage of historical data. (b) Kinetics of oxygenation of the water body using the pumping-Venturi tube system.

4. Conclusions

This work reports, via simulation, the derivation of an equation, based on historical data, hydraulic relations, and the equation for mass conservation. The experimental validation with an air compressor and a Venturi tube was also carried out. To determine if the proposed model is capable of describing the experimental data obtained from an air injection compressor to the water body, a water pumping and air capture system was built through a Venturi tube from which oxygenation kinetics of the water body was obtained. In both cases, it was observed that the proposed mathematical model is capable of predicting the saturation DO concentration inside a body of water with a measurement error of 1 mg/lt. This equation could be implemented into physical aeration systems to reduce the associated energy consumption in tilapia farms. Therefore, it is recommended to use this mathematical model in water growth ponds whose interior species have been cataloged as extremophiles.

As future work, it is proposed to minimize the measurement error of the model through the identification and incorporation of parameters that can be easily measured by the operator of an aquaculture farm, or through calculating a correction factor that minimizes the error obtained in the factorial regressions.

Acknowledgments
The authors of this work want to thank to the laboratory in applied research and innovation of the Technological Institute of Colima for the resources and space given for the development of this project.

References
[1] Xamán J 2016 *Dinámica de Fluidos Computacional para Ingenieros* (México: Palibrio)
[2] Mott R 2006 *Mecánica de Fluidos, sexta edición* (México: Pearson Educación)
[3] Pazmino J, García A, Fernández J, Cisneros C 2017 Diseño y construcción de un prototipo de medidor de Venturi con fines didácticos y aplicación a la ingeniería agronómica *Polo del Conocimiento 2.5* **1** 1033
[4] García A 2011 *Ventilación y Oxigenación en Situaciones de Urgencia* (Madrid: Asociación Española de Pediatría de Atención Primaria)
[5] Atoche E, Yrs B 2021 *Diseño de Box-Behnken en la Eficiencia del Sistema Venturi de Nanomicroburbujeo de Oxígeno para la Remoción de Materia Orgánica y Microbiológicas en Aguas Residuales Domesticas* (Perú: Universidad Peruana Unión)
[6] Astuhuaman A 2009 *Estudio de Parámetros en la Lixiviación en Pilas para Optimizar la Extracción de Oro* (Perú: Universidad Nacional del Centro de Perú)
[7] Comisión Nacional de Acuacultura y Pesca (CONAPESCA) 2018 *Buenas Perspectivas para la Exportación de Pescados y Mariscos Mexicanos* (México: Comisión Nacional de Acuacultura y Pesca)
[8] Boyd C 1990 *Quality in Ponds for Aquaculture, Alabama* (Alabama: Auburn University)
[9] Huaito S, Rogger M 2017 *Control Avanzado Sobre un Sistema de Aireación Durante el Cultivo de Langostinos (Litopenaeus Vannamei), Usando Modelos Matemáticos para la Predicción de Oxígeno Disuelto en Acuicultura* (Perú: Universidad de Piura)
[10] Food and Agriculture Organization of the United Nations (FAO) 2010 *Documento Técnico de Pesca* (United States of America: Food and Agriculture Organization of the United Nations)
[11] Defe G, Antonio A 2018 Multi-parameter Water Quality Monitoring Device for Grouper Aquaculture *IEEE 10th International Conference on Humanoid, Nanotechnology, Information Technology, Communication and Control, Environment and Management* (Baltimore: IEEE)
[12] Mena A 2001 *Relación entre la Gasometría y las Variables Productivas de Tilapia Roja Oreochromis Niloticus Durante la Adaptación y Cultivo a Diferentes Salinidades* (Colima: Posgrado Interistitucional en Ciencias Pecuarias)
[13] American Society of Civil Engineers (ACSE) 2006 *Measurement of Oxygen Transfer in Clean Water* (United States of America: American Society of Civil Engineers)
[14] Cheng X, Xie Y, Zhu D, Xie J 2019 Modeling re-oxygenation performance of fine-bubble–diffusing aeration system in aquaculture ponds *Aquaculture International* **27** 1353

Journal of Physics: Conference Series **2153** (2022) 012018 doi:10.1088/1742-6596/2153/1/012018

Simulations of an impulsive model for the growth of fruit trees

E Duque-Marín[1], A Rojas-Palma[1], and M Carrasco-Benavides[2]

[1] Facultad de Ciencias Básicas, Universidad Católica del Maule, Talca, Chile

[2] Facultad de Ciencias Agrarias y Forestales, Universidad Católica del Maule, Curicó, Chile

E-mail: `fedwer@gmail.com`, `amrojas@ucm.cl`

Abstract. Mediterranean agricultural systems have been severely affected because of the decrease in rainfall and more frequent and severe droughts due to the global warming phenomenon. The current and future scenario of water deficit could have a negative effect on the growth and development rates of the fruit trees, reflected in the drop of production. To help to face this problem, this work presents a mathematical simulation model of fruit growth with two-time scales: a continuous scale that governs the dynamics of fruit growth and a discrete scale representing the period of time in which the system is intervened with irrigation supply. The results obtained in the simulations allow us to describe and understand the physical phenomena involved in the growth dynamics of fruit trees. In addition, show the importance of the water resource for the growth and development of fruit trees; therefore, a scenario of water deficit would compromise the production and existence of fruit trees.

1. Introduction

The scarcity of water generated by climate change is modifying the production and conservation of agricultural systems. The increase in global temperature [1, 2], along with diminishing water supply, is altering in a negative way the growth and production cycles of fruit trees growing under a Mediterranean-type climate. In Mediterranean semi-arid production conditions, all fruit trees are produced under irrigation. Irrigation is defined as "the application of water to a soil profile, in a timely and uniform manner, to replace the water consumed by the crop between two consecutive irrigations" [3].

Thus, especially in semi-arid areas, the reduction in the water supply for irrigation and the increase of dry and hot weather conditions, are dramatically increasing the risk of fruit production because of their effects on the growth and development rates [4, 5]. Consequently, understanding the interactions between irrigation and fruit trees production is essential to the fruit growers to take decisions to sustainably manage the fruit trees production, especially right now where environmental conditions are adverse.

Crop growth modeling has been developed widely with engineering-type models, supported by biophysical sub-models based, for example, on the second Newton's law, and Darcy's laws, that combined with a numeric fit, provide essential information for support the water management at field [6–8]. Some models of this type are crop yield prediction models [9–11], models of crop response to environmental factors [12, 13] and simulation models such as world food studies (WOFOST), used in [14] to simulate the growth of jujube under different irrigation treatments.

Engineering models are characterized by using statistical tools and computer programs to find trends and validate field data. Scientific models are also evidenced that contribute to understanding the response of crops to environmental factors. Among them, those exposed in [15] stand out. In [16] a study of the growth dynamics of fruit trees under a water scarcity scenario is presented.

The construction of a mathematical model implies analysis and understanding of the problem to be modeled, identifying the variables and parameters involved, and the relationships between them, in order to reflect reality simply but precisely. However, as far as the authors are aware, there is no evidence of a dynamic model that describes the interaction between the 3 variables: energy, biomass of the fruit tree, and water, incorporating the supply of irrigation. This article presents the simulation of an impulsive model for the growth of fruit trees with different irrigation strategies.

2. Mathematical modeling

The models for the growth of fruit trees involve the interaction between different variables such as genotype, crop management, type of soil, salinity, environmental factors, water, energy, among other variables [15–19]. For the construction of an impulsive mathematical model for the growth of fruit trees, we consider the following assumptions:

(i) The growth dynamics is governed by an interaction between the variables energy, water, vegetative growth.
(ii) The fruit tree responds instantly to the irrigation application.
(iii) We assumed an adult fruit tree (older than 5 years) and a suitable soil surface for the growth of the fruit tree.
(iv) Ideal agronomic management conditions.
(v) Optimal environmental conditions are assumed, that is, the energy of the system is constant.

The parameters that intervene in the dynamics: water, energy and vegetative growth in our model are summarized in Table 1.

Table 1. Parameters used in the fruit tree growth model.

Parameter	Meaning
q	Accumulated energy constant
r	Fruit trees intrinsic growth rate
N	Fruit carrying capacity
I	Irrigation water amount
β	Evapotranspiration rate
γ	Photosynthetic contribution rate
ω	Mortality rate of fruit trees

We denote $E = E(t)$ the solar radiation at time t, $W = W(t)$ the water amount in the soil in time t and $C = C(t)$ the fruit biomass concentration in time t, both state variables. Under assumption (iv), the fruit growth dynamics is reduced to two-state variables W and C. The variation of the amount of water in the system with respect to the time is denoted by $W'(t)$. When considering a scenario without precipitation, this is $p = 0$, the water variation presents output from the system due to environmental effects (evapotranspiration) at a rate β. At the same time, there is another water outlet from the system corresponding to promoting the growth of the fruit tree at a rate r in the sistem.

The variation in biomass growth in the system with respect to the time is denoted by $C'(t)$ corresponds to the input by water-biomass growth interaction at a rate r, in addition, this variation is also positively affected due to the water-energy-biomass growth interaction at a rate γ. in addition, there is an exit to an ω rate corresponding to loss or natural death of the crop. This behavior applies to a continuous time scale (long duration) that governs the growth dynamics of the fruit tree. However, this model also presents a short time scale (pulse) in discrete time that represents the events in which water enters the system through irrigation supply.

The dynamic relationship between the variables energy, amount of water, and concentration of biomass of the fruit tree is represented in Figure 1; where: E = Energy; W = Water amount; C = Fruit biomass concentration; r = Fruit trees intrinsic growth rate; N = Fruit carrying capacity; β = Evapotranspiration rate; γ = Photosynthetic contribution rate; ω = Mortality rate of fruit trees; p = Rainfall rate; I = Irrigation amount, adapted from [16].

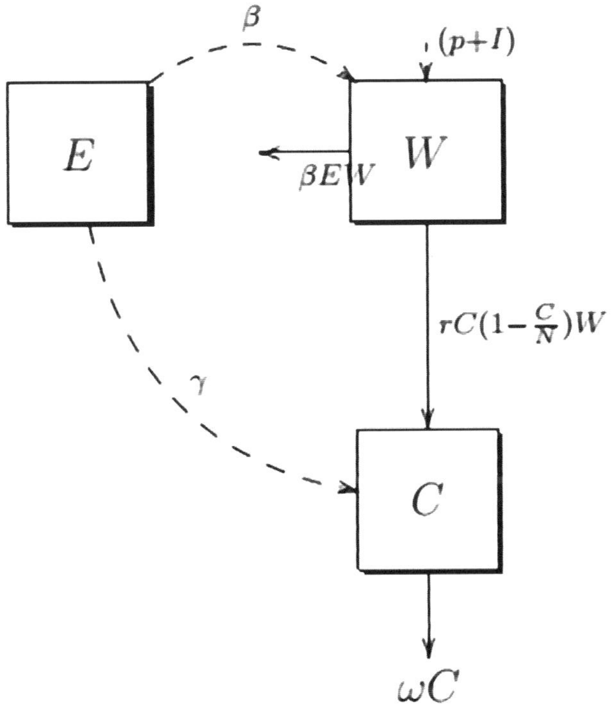

Figure 1. Compartment diagram of dynamics.

Therefore, a model is proposed that represents the growth dynamics of fruit trees, where the supply of irrigation is evidenced in the form of a pulse. This model can be described by the impulsive differential system of equations, Equation (1).

$$\begin{cases} W'(t) &= -\beta q W(t) - rC(t)(1 - \frac{C(t)}{N})W(t) \\ C'(t) &= rC(t)(1 - \frac{C(t)}{N})W(t) + \frac{\gamma q C(t) W(t)}{(C+1)(W+1)} - \omega C(t) \quad t \neq nT. \\ \Delta W(nT) &= I \\ \Delta C(nT) &= 0 \qquad t = nT. \end{cases} \tag{1}$$

3. Results

In the different simulations, the state variables are represented. In Figure 2, the amount of water in the system (blue line) shows the behavior of this variable with the implementation of an irrigation strategy that considers the application of the same amount of irrigation water in 8 regular periods during the season. Figure 2 shows how the amount of water increases in the moments $t = nt$ in which the irrigation intervention is carried out, subsequently the loss of this is observed due to the effects of evapotranspiration and the demand of the fruit tree. For its part, the growth curve of the biomass of the fruit tree (red segmented line) shows the response that this variable has to the intervention of the irrigation system.

Figure 2 shows a growth of the biomass of the fruit tree as a response to the availability of water, showing greater growth in those stages in which irrigation has been applied, that is, in the moments in which there is the availability of water resources for satisfying the demand of the fruit tree. In Figure 2 the blue line corresponds to the amount of water in the soil ($W(t)$), and the segmented red line to the biomass concentration of the fruit tree ($C(t)$). The initial conditions are $W(0) = 0.6$, $C(0) = 1$ and parameter values $r = 0.043$, $q = 0.5$, $N = 3000$, $I = 1$, $\beta = 0.06$, $\gamma = 0.001$, $\omega = 0.00001$.

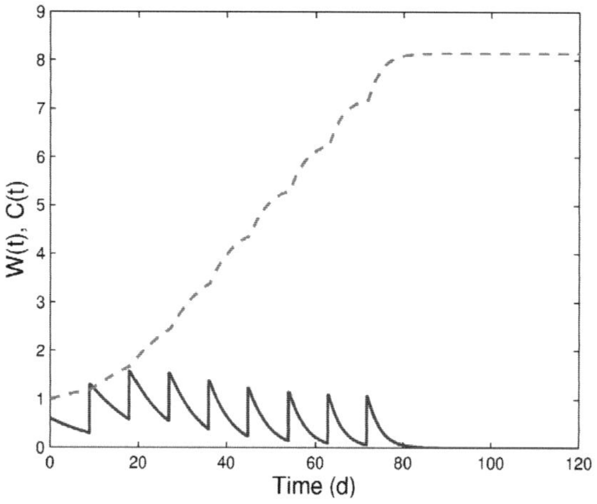

Figure 2. Simulations of the soil water amount (W) and dynamic of growth of the fruit tree biomass (C).

Figure 3 shows the response of biomass growth to different environmental conditions. In this simulation, the irrigation strategy implemented in Figure 2 is preserved. However, the response of the growth of the biomass of the fruit tree is simulated for different values of accumulated energy within the system. The established values are $q = 0.1$, $q = 0.5$, $q = 1$, blue, red and green line respectively. Here, the initial condition is $C(0) = 1$ and parameter values $r = 0.043$, $N = 3000$, $I = 1$, $\beta = 0.06$, $\gamma = 0.001$, $\omega = 0.00001$.

In this regard, the results presented in Figure 3 show how environmental conditions affect the growth and production of fruit trees. The blue line represents the response of the fruit tree to small values of accumulated energy, that is, it presents favorable environmental conditions for the growth and development of the fruit tree.

On the other hand, when considering an average value of accumulated energy (red line), a growth of the fruit tree is evident, however, a maximum yield is not presented. The green line presents an adverse condition in which a loss of crop yield is shown because it is a scenario with an irrigation strategy that does not satisfy the water demands of the fruit tree generated by strong environmental conditions.

The amount of water in the system is presented in Figure 4(a), which shows two different irrigation strategy scenarios. The first scenario (blue line) considers the application of the same amount of water for 8 regular periods of time. The (red line) represents the second scenario in which a fixed amount I of irrigation water is applied in regular longer periods. Both scenarios show the decrease of this variable over time, with the second scenario presenting a more pronounced loss. Here, the initial conditions are $W(0) = 0.6$, $C(0) = 1$ and parameter values $r = 0.043$, $q = 0.5$, $N = 3000$, $I = 1$, $\beta = 0.06$, $\gamma = 0.001$, $\omega = 0.00001$. In addition, the biomass of the fruit tree is represented in Figure 4(b), this shows the response of the fruit tree to the different irrigation strategies. With $W(0) = 0.6, C(0) = 1$ and parameter values $r = 0.043$, $q = 0.5$, $N = 3000$, $I = 1$, $\beta = 0.06$, $\gamma = 0.001$, $\omega = 0.00001$.

In Figure 4(b) the response of fruit growth (blue line) to the first scenario shows that the application of irrigation in regular periods of time enhances the crop, however, there is evidence of alterations in fruit growth, more appreciable in the latter irrigation applications, this occurs because the water demand is not satisfied. The response of fruit growth to the second irrigation strategy (red line) shows that the application of irrigation between long periods of time extends the water requirement of the fruit tree and affects growth by decreasing its biomass.

These results agree with those presented in [20] who used the farming systems model agricultural production systems simulator (APSIM), for simulating and validating crop yield and field water balance of the wheat-corn rotation in the North China plain. The result indicates that an Increased amount of irrigation led to increased crop yield. The simulations presented contribute to the knowledge and description of the fruit growth phenomenon. This methodology broadens the approach to the study of crop modeling, especially fruit trees whose complex dynamics have been studied from physical aspects.

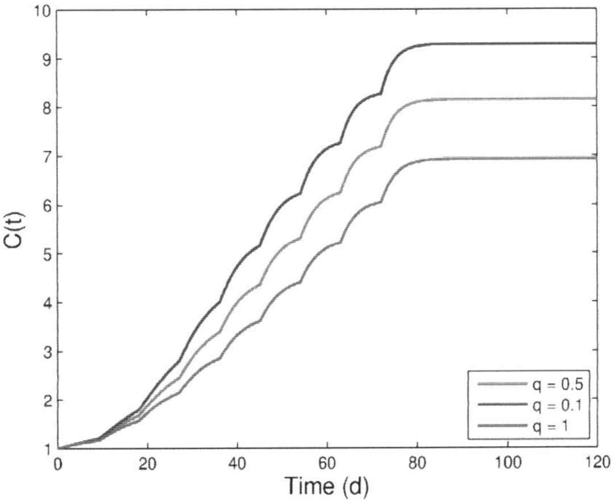

Figure 3. Simulations of dynamic of growth of the fruit tree biomass C.

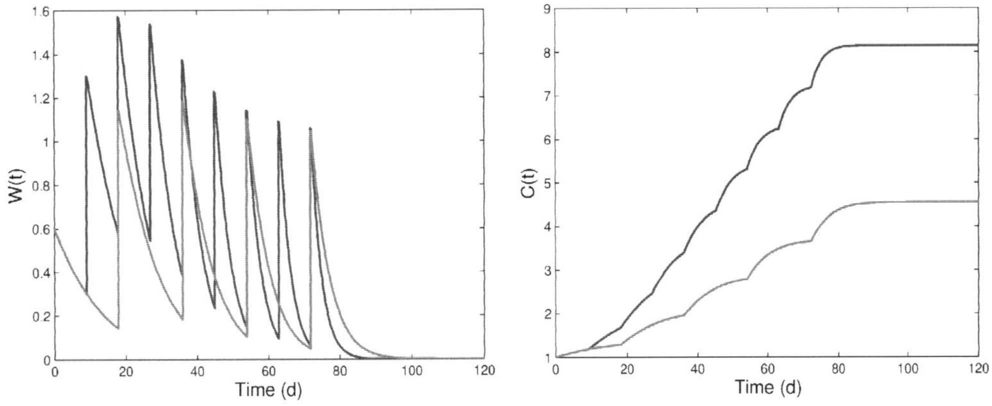

(a) (b)

Figure 4. Simulations of: (a) the soil water amount $W(t)$, and (b) the dynamic of growth of the fruit tree biomass $C(t)$.

4. Conclusion

In this work, simulations of an impulsive system of nonlinear differential equations were presented to describe the response of fruit tree growth to the application of water by irrigation. The increase in temperature, the irrigation times and the amount of irrigation to be supplied, are some of the factors that were considered in the different simulations, with the aim of showing how the variation of any of them can compromise the growth of the fruit trees. Therefore, the different simulated scenarios respond to concerns that must be addressed when implementing an irrigation strategy that contributes to improving the production of fruit trees.

The results of this work showed the positive effects of the implementation of adequate irrigation scheduling, which satisfies the water demands of the fruit trees considering the adverse climatic conditions that arise. The novelty of this work is that it offers an easy-to-use mathematical qualitative model to numerical simulate the effects of different water supply scenarios on fruit production as an alternative to other complex engineering models.

Therefore, we can conclude that in the current scenarios of decreased rainfall and frequent drought, the application of irrigation water becomes a prevailing need for the production, conservation, and persistence of fruit trees. However, it is necessary to continue with the study of the growth dynamics of fruit trees using impulsive models in which the supply of irrigation variables overtime is considered and to analyze how this affects the development and productivity of fruit trees.

Acknowledgments

The present research was partially supported by the Ph.D. scholarship of the Universidad Católica del Maule, and the Chilean government through the Agencia Nacional de Investigación y Desarrollo (ANID) throughout the Programa FONDECYT Iniciación en la Investigación, año 2017" (grant No. 11170323).

References

[1] Rueda V O M 2004 El cambio climático global: comprender el problema *Cambio Climático: Una Visión desde México* (México: Secretaría de Medio Ambiente y recursos Naturales Instituto Nacional de Ecología)
[2] Lobell D B, Field C B 2007 *Environmental Research Letters* **2(1)** 014002
[3] Gurovich L A 1985 *Fundamentos y Diseño de Sistemas de Riego, No 59* (Costa Rica: Instituto Interamericano de Cooperación para la Agricultura)
[4] Ojeda-Bustamante W, Sifuentes-Ibarra E, Íñiguez-Covarrubias M, Montero-Martínez M J 2011 *Agrociencia* **45(1)** 1
[5] Diaz J R, Weatherhead E K, Knox J W, Camacho E 2007 *Regional Environmental Change* **7(3)** 149
[6] Steduto P, Hsiao T C, Raes D, Fereres E 2009 *Agronomy Journal* **101(3)** 426
[7] Passioura J B 1996 *Agronomy Journal* **88(5)** 690
[8] Waller P, Yitayew M 2015 *Irrigation and Drainage Engineering* (New York: Springer)
[9] Palosuo T, *et al.* 2011 *European Journal of Agronomy* **35(3)** 103
[10] de Wit A, *et al.* 2010 *Climate Research* **44(1)** 41
[11] Bussay A, van der Velde M, Fumagalli D, Seguini L 2015 *Agricultural Systems* **141** 94
[12] Asseng S, *et al.* 2013 *Nature Climate Change* **3(9)** 827
[13] Todorovic M, Albrizio R, Zivotic L, Saab M T A, Stöckle C, Steduto P 2009 *Agronomy Journal* **101(3)** 509
[14] Bai T, Zhang N, Chen Y, Mercatoris B 2019 *Sustainability* **11(5)** 1466
[15] Thornley J H, Johnson I R 1990 *Plant and Crop Modelling* (New York: Oxford)
[16] Duque-Marín E, Rojas-Palma A, Carrasco-Benavides M 2021 *Journal of Physics: Conference Series* **2046(1)** 012017:1
[17] Keating B A, Thorburn P J 2018 *European Journal of Agronomy* **100** 163
[18] Monteith J L 1996 *Agronomy Journal* **88(5)** 695
[19] Ewert F, *et al.* 2015 *Environmental Modelling & Software* **72** 287
[20] Chen C, Wang E, Yu Q 2010 *Agricultural Water Management* **97(8)** 1175

Biomechanical analysis of grip strength in upper limb rehabilitation

V K Hernández Vergel[1], R Prada Núñez[2], and C A Hernández Suárez[2]
[1] Facultad de Ciencias Médicas y de la Salud, Universidad de Santander, San José de Cúcuta, Colombia
[2] Facultad de Educación, Artes y Humanidades, Universidad Francisco de Paula Santander, San José de Cúcuta, Colombia

E-mail: raulprada@ufps.edu.co

Abstract. This research is based on biomechanics as a science that involves concepts of engineering, mechanics, physic, anatomy, physiology, and many others, to study the human body with the desire to solve certain problems that may affect the performance of an individual in their work or personal level. This work is an investigative process in these areas of scientific and applied disciplines, in which the attention is focused on the hand as a valuable tool for the occupational performance of the human being, since through it is possible to touch, move, grasp, or manipulate objects. Injuries to this limb may be due to various causes, which require complex surgeries and long periods of rehabilitation to be reversed. This research highlights the importance of certain physical concepts that must be understood by the rehabilitation expert in order not to affect the surgery and thus guarantee the maximum functionality of the patient at the end of the recovery cycle.

1. Introduction

The hand is the element of the human body that interacts with the environment either to move, touch, manipulate or grasp objects, which makes it a necessary limb for daily life and work due to its functionality and ability to adapt to the shape of objects. As a factor that decreases work productivity or functionality, when it is affected by injuries, the maximum force that would be exerted under normal conditions is diminished.

However, the correct diagnosis and treatment of injuries has its own philosophy, since each injury compromises the biomechanics of the tendon, which motivates its in-depth study. The extensor tendon can be injured in different ways, generally the most common injuries are caused by lacerations, crushing, and forced flexion of the joint which, when extended, can cause laceration of the tendon insertion with or without fracture of the phalanx. Biomechanical analysis to characterize the force exerted by the hands is performed by means of readings obtained from dynamometers, either mechanical or hydraulic, where the results correspond to the force exerted by the whole hand.

Several previous clinical studies [1,2] have provided valuable information on this type of injury, for example it is known that: (a) extensor tendons in all areas tolerate controlled active mobility; (b) breaching and rupture are rarely a problem in carefully applied postoperative regimens where forces and excursion are controlled; (c) more digital joint mobility can probably be allowed than previously thought possible; (d) wrist position is critical in decreasing restrictive forces on the flexor system and is a factor in the true tendon excursion obtained with digital mobility; (e) those tendons have probably been actively moved, all within the confines of their dynamic splints; and (f) early referral to therapy with attention to splint geometry and applied stress are critical variables in achieving results [3].

Therefore, when designing a splint for the rehabilitation of the extensor tendons of the hand, the aim is to use mechanical means to allow the movement of the hand, generating faster and more effective treatment responses and thereby reducing the biomechanical, physiological, psychological, and socio-occupational impact.

Biomechanics studies the models, phenomena and laws governing the movement and balance of the human body. It is based on the biomedical sciences, which use knowledge of mechanics, anatomy, physiology, and other disciplines to study the behavior of the human body and to solve problems arising from the various conditions to which it may be subjected. Studies based on the human body and the human-machine relationship have achieved the restitution or replacement of damaged functions and structures, which has encouraged the construction of instruments for therapeutic and diagnostic purposes such as prostheses or devices that replace a biological function [4].

Most biomechanical studies focus on the hand, due to the importance it represents in mobility and the ability to manipulate and hold objects in daily tasks. For this reason, carrying out an in-depth study of this upper limb has become a necessary activity in research, either to replace it by means of prostheses or for rehabilitation in people with hand impairments. This study will serve as a support for health personnel. For this reason, a characteristic study of the forces generated by the hand and how it is affected after an injury after the rehabilitation process has been completed is proposed. The objective of any rehabilitation cycle is to guarantee the maximum functionality of the individual.

2. Method

To determine the degree of effectiveness and efficiency of rehabilitation treatments for extensor tendon injuries from zone IV to zone VIII, a descriptive comparative study was carried out to identify the most appropriate recovery protocol, considering the type of splint used in the recovery process as an independent variable and the intensity of the strength reached by the affected upper limb at the end of the rehabilitation process as a dependent variable.

The application of the protocols to determine the effectiveness and efficiency in the recovery process of the extensor tendons was carried out with patients who underwent surgery at the University Hospital in the city of San José de Cúcuta, Colombia, and who subsequently went to a rehabilitation center. The sample was non-probabilistic and intentional, since the inclusion criteria applied were the injury in the study zones and that they were able to complete the initial rehabilitation process (six weeks). A sample of six patients was selected, distributed equally between the traditional and early mobility controlled active movement (ICAM) treatment protocols.

For the pressure measurement, certain conditions were determined according to the American Association of Hand Therapists [5] emphasizing that the person should be in a seated posture, with spine aligned, shoulders adducted and without rotation, the elbow forming a 90° angle to the side of the body, together with the forearm and wrist in a non-influential position, *i.e.*, in a neutral position.

We had the support of a materials supply corporation, which provided the necessary resources to produce the splints, while the rehabilitation center has the Jamar dynamometer [6] and the Martin Vigorimeter [7], both instruments used to measure grip force and pressure.

3. Results and discussion

Regarding the demographic characteristics of the people who participated in the research, their ages range between 15 and 60 years, with an average age of 38.9 years. There is a predominance of men, whose daily activities are mainly concentrated in mining and agriculture, with a predominance in the municipalities of the Catatumbo region, Colombia. The patients to whom the traditional process was applied presented lesions in zones IV and V with a predominance of the third finger of the left hand, but predominantly on the right side; while the patients to whom the ICAM protocol was applied presented lesions in zone V, with 66.7% on the right side and the remaining percentage on the left hand.

Journal of Physics: Conference Series **2153** (2022) 012019 doi:10.1088/1742-6596/2153/1/012019

3.1. Biomechanical analysis of grip force

The hand corresponds to the final section in the mechanical line of levers starting with the shoulder, passing through the elbow, the wrist and ending with the wrist, characterized by operating in various planes, which allows the hand to move within a wide space and enables it to reach all parts of the body with great ease.

The grip strength of the hand corresponds to the force used to suspend objects in the air or to squeeze them. This requires the correct position of the wrist to avoid the development of so-called cumulative trauma disorders [8]. The grip force can be squeezing, clamping, and supporting. Then the biomechanical characterization of the grip involves the determination of the different kinematic and dynamic variables that affect the different phases of the grip: transport, grip formation and manipulation, as well as their relationship with the anthropometric characteristics of the subject and those of the manipulated object and the manipulation task [9].

Some key parameters that contribute to the characterization of the contact between an object and the hand are the areas of the hand in contact and the pressure experienced on it. This information is necessary as input data in the generation of biomechanical models of the hand to investigate the muscular efforts that act in this grip. Functional aspects of the physiology of the hand (muscles and tendons) are implicit in the hand's grip action or hand movement, among which the deep flexor of the fingers is the generator of pressure given its physical characteristics [10], in support of other muscles.

In the research of [9], an analysis of the forces generated internally with angles of 120° and 90° of flexion of a proximal interphalangeal joint is carried out and it is concluded that to generate an optimal grip where the forces generated will be maximum with respect to the tension in the flexor tendons, it is required that the fingers are as closed as possible (see Figure 1). It is in this scenario where the rehabilitation process plays a key role. The resultant forces in each force system are obtained from the application of Equation (1).

$$F_R = \sqrt{F_1^2 + F_2^2 + 2 * F_1 * F_2 * \text{Cos}\theta}. \tag{1}$$

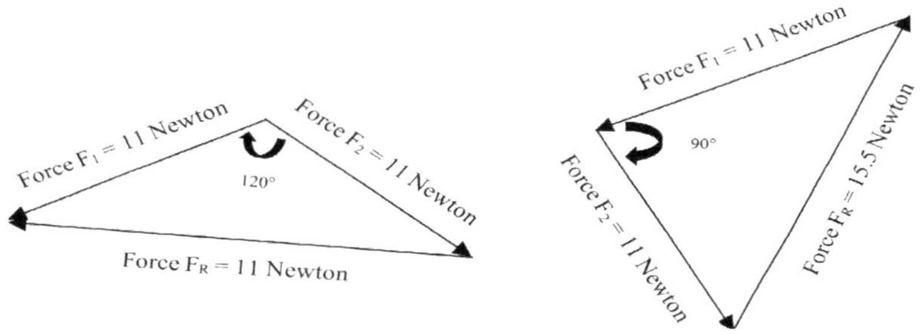

Figure 1. Magnitude of forces generated by the hand and their relation to the flexion position at the interphalangeal joint.

During the rehabilitation process, use was made of the dynamometer (cylindrical palm pressure) and the vigorimeter (spherical palm pressure) in order to consider, as a secondary objective, whether the shape influenced the measurement due to the comfort of the affected limb, as described by [7] where it is highlighted that the palm pressures are those in which the fingers and palm of the hand are involved, corresponding to a force pressure with the purpose of taking bulky and heavy objects [7], as shown in Figure 2.

Table 1 presents the measurements taken at the end of each week of treatment using the cylindrical palmar pressure measurement device (with force recording in kilograms) once the patient was able to start the rehabilitation process. Two measurements were taken for each patient with a five-minute recovery interval in between. In this way, six measurements per week were obtained, to report some pressure descriptions such as average, standard deviation (SD) and the lower and upper limits of the 95% confidence interval for each of the two treatments under evaluation.

The results offered by the patients who underwent the ICAM early mobility treatment show better results in terms of the recovery of their affected hand, compared to the traditional process, so the difference is statistically significant. Although it is worth clarifying that, in both cases, the strength measurements correspond to the measurements that would be reported by an infant of no more than 13 years of age [7].

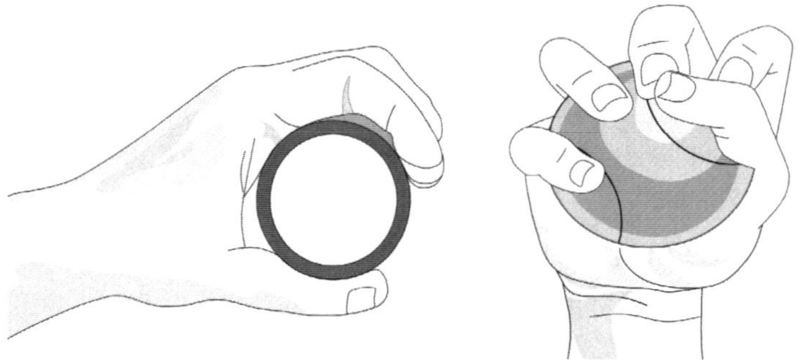

Figure 2. Palmar pressure (a) cylindrical and (b) spherical. Adapted from [7].

Table 1. Comparative descriptive data according to rehabilitation treatment using the Jamar dynamometer.

Week	Traditional method				ICAM protocol			
	Average	SD	Lower limit	Upper limit	Average	SD	Lower limit	Upper limit
1	9.83	1.24	8.84	10.82	12.52	1.13	11.62	13.42
2	10.99	1.78	9.57	12.41	13.24	1.68	11.90	14.58
3	12.84	1.94	11.29	14.39	15.46	2.11	13.77	17.15
4	14.86	2.33	13.00	16.72	16.77	2.25	14.97	18.57
5	15.65	2.97	13.27	18.03	17.48	2.78	15.26	19.70
6	17.23	3.11	14.74	19.72	21.23	3.31	18.58	23.88

Table 2 presents the measurements taken at the end of each week of treatment using the spherical palmar pressure measurement device (with force recording in bar) once the patient was able to start the rehabilitation process. Two measurements were taken for each patient with a five-minute recovery interval in between. Six measurements per week were obtained, to report some descriptive pressure data such as average, standard deviation (SD) and the lower and upper limits of the 95% confidence interval for each of the two treatments under evaluation.

The results obtained from the patient measurements show slightly higher values for those who received the ICAM early mobility treatment, but these are not statistically significant, *i.e.*, there are no differences in the evolution of the two treatments under this measurement tool. It continues to be highlighted that the measurements report values that would correspond to an infant of no more than 13 years of age without any type of affectation in their hand [11].

There are few works that report the parameters of grip strength in people according to their age, for example, in [11,12] fist strength parameters are reported in healthy people of working age, while in [13] parameters are defined between 6 and 18 years of age, but in the three investigations the use of the dynamometer alone is resorted to. While the characteristics of the research developed by [7] is more in line with the characteristics of this research in terms of the use of various instruments for the measurement of the pressure force that vary in the way the hand is gripped to perform the measurement, which becomes a relevant feature given the characteristics of affectation of the informants in this research.

It should be noted that for the purposes of this discussion of results, it is compared with the parameters provided in [7] who estimate the parameters of normal strength in children and young people. This situation becomes an obstacle that makes it more complex to determine both the degree of affectation of an injury and the level of effectiveness of the rehabilitation process.

The differences found between the rehabilitation treatments with the use of the Jamar dynamometer are based on the way in which the person performs the grip, favoring those patients who have been rehabilitated with the ICAM treatment, because, as concluded in [9], the optimal grip where maximum forces are reached with respect to the tension in the flexor tendons occurs when the fingers are as closed as possible.

Table 2. Comparative descriptive data according to rehabilitation treatment using the Martin Vigorimeter.

Week	Traditional Method				ICAM protocol			
	Average	SD	Lower limit	Upper limit	Average	SD	Lower limit	Upper limit
1	0.273	0.045	0.24	0.31	0.368	0.051	0.33	0.41
2	0.347	0.073	0.29	0.41	0.375	0.057	0.33	0.42
3	0.389	0.074	0.33	0.45	0.392	0.063	0.34	0.44
4	0.396	0.077	0.33	0.46	0.406	0.063	0.36	0.46
5	0.413	0.078	0.35	0.48	0.421	0.065	0.37	0.47
6	0.422	0.081	0.36	0.49	0.448	0.069	0.39	0.50

4. Conclusions

After the execution of this research, several aspects can be highlighted: is evidence of the interdisciplinary knowledge that generates antecedents of work in science, technology, engineering, the arts, and mathematics, where the success of a rehabilitation process is clearly influenced by the therapist's knowledge of physics concepts, such as force, tension, pressure or pulleys that are immersed in the biomechanical functioning process of the hand; the evaluation of the effect of the rehabilitation process is clearly influenced by the instrument used for force measurement, given the particular connotation of this study, where the informants were people who had suffered hand injuries between zones IV to VIII, so the position of the fingers becomes a differentiating factor; finally, it is important to continue extending the results in this line of work, including people whose work activities are different from mining or agriculture, which correspond to the occupations of these participants, since clearly and after reviewing the literature, the patient's work activity influences their rehabilitation process; it is in this scenario where the concepts of Physics provide valuable guidance to the rehabilitation professional in order to guarantee intervention protocols that contribute to the functional recovery of the affected limb.

References

[1] Gil L, Beltrán J A, Barrios C 2014 Lesiones deportivas de la muñeca y mano *Arch Med Deporte* **31(159)** 41
[2] Delgado A D, Ienéltiri T 2001 Lesiones de la mano en urgencias *Med Integral* **38(8)** 363
[3] Masméjean E, Le Bellec Y L, Alnot J Y 2000 Lesiones traumáticas de los tendones extensores de la mano *EMC - Aparato Locomotor* **33(4)** 1
[4] Espinosa A, Serrano L, Zárate J G, Romo R 2009 Resultados de la fijación con anclas en lesiones de ligamentos y tendones de la mano y la muñeca *Acta Ortopédica Mexicana* **23(2)** 70
[5] Moreira D, Aiza R R, De Gogoy J R, Do Nascimento A 2003 Abordagem sobre prensao palmar utilizando o dinamómetro JAMAR: uma revisao de literatura *Brazilian Journal of Science and Movement* **11(2)** 95
[6] Hincapié O 2007 Elaboración de estándares de fuerza de agarre en individuos sanos entre 20 y 70 años residentes en la localidad de Usaquén, Bogotá *Revista Colombiana de Rehabilitación* **6(6)** 5
[7] Escalona P, Naranjo J, Lagos V, Solís F 2009 Parámetros de normalidad en fuerzas de prensión de mano en sujetos de ambos sexos de 7 a 17 años de edad *Revista Chilena de Pediatría* **80(5)** 435
[8] Mastalerz A, Nowak E, Palczewska I, Kalka E 2009 Maximal grip force during holding a cylindrical handle with different diameters *Human Movement* **10(1)** 26

[9] Escobar J H, Quiroz R A 2014 *Análisis Biomecánico de las Fuerzas de Agarre de la Mano Mediante un Dispositivo que Simula las Lesiones de Amputación de Falanges* (Bucaramanga: Universidad Industrial de Santander)

[10] Nordin M, Frankel V H 2004 Biomecánica de la muñeca y de la mano *Biomecánica Básica del Sistema Musculoesquelético* ed Nordin M, Frankel V H (Madrid: Mc Graw Hill)

[11] Mahn J, Romero C 2005 *Evaluación de la Fuerza de Puño en Sujetos Adultos Sanos Mayores de 20 años de la Región Metropolitana* (Santiago de Chile: Universidad de Chile)

[12] Bustos-Viviescas B J, Acevedo-Mindiola A A, Lozano-Zapata R E 2019 Valores de fuerza prensil de mano en sujetos aparentemente sanos de la ciudad de Cúcuta, Colombia *MedUNAB* **21(3)** 363

[13] Marrodán M S, Romero J F, Moreno S, Mesa M S, Cabañas M D, Pacheco J L, González-Montero M 2009 Dinamometría en niños y jóvenes de entre 6 y 18 años: valores de referencia, asociación con tamaño y composición corporal *Anales de Pediatría* **70(4)** 340

Newton's law learning assessment: an experience with high school students

R Prada Nuñez[1], **C A Hernández-Suarez**[1], **and A A Gamboa Suarez**[1]
[1] Universidad Francisco de Paula Santander, San José de Cúcuta, Colombia

E-mail: cesaraugusto@ufps.edu.co

Abstract. The Force Concepts Inventory is a test to determine the level of conceptual knowledge of students about mechanical physics and to evaluate the effectiveness of different teaching strategies on the conceptual component of learning. The test is applied with the purpose of knowing the level of conceptualization of the students of a Physics subject course. The results of the pre-test made it possible to find out the level of conceptualization that the students possessed and provided information for the development of workshops based on real physical situations that required the elaboration of force diagrams. The results of the post-test allowed estimating Hake's learning and showed evidence about the conceptual evolution of the students and information to develop future teaching activities on Newton's laws.

1. Introduction

In mechanical physics, the concept of force is fundamental in academic programs in science and engineering, so students (who come from pre-university education), need to conceptualize it to apply Newton's laws in the resolution of different concrete problem situations. One of the curricular orientations for the teaching of natural sciences in secondary education (by the "Ministerio de Educación Nacional, Colombia"), is related to the theoretical and practical foundations of classical mechanics, from the movement of bodies and their interactions, the concept of force, work, and energy, supported by mathematical modelling, for the explanation of situations in nature [1,2].

Therefore, the students' learning, before focusing on algorithmic and algebraic procedures, must be oriented towards a conceptual domain that allows them to understand the laws and the explanation of physical phenomena, since, when approaching the solution of a problem, it is decontextualized from the conceptual arguments of physics. This adds to the inherent difficulty of concepts and their relationships to be understood; acceleration and force, equivalence between rest state and movement with constant speed, inertial and non-inertial systems, among others [3].

One of the difficulties that hinder the learning of mechanical physics is related to the erroneous preconceptions of the students [4], that are born from the daily experience that led to internalize incorrect relations between different physical magnitudes, for example, to use the term force in a variety of contexts, using vague and ambiguous associations [5], which are difficult to modify in structural terms.

The personal construct of students' previous ideas are representational schemes that do not model adequate scientific conceptions, thus becoming an obstacle that does not favor conceptual change despite teaching attempts [6]. Therefore, helping a conceptual change requires teaching activities that start from what the student knows, in a social teaching context [7,8], that favor the clear, stable, and organized reconstruction of knowledge with which they can face and solve diverse situations and, at the same time, acquire knowledge that demands a higher level of abstraction within the same field.

The intervention developed in this study that has as its axis of analysis the consistent representation of the interactions between systems through the so-called force diagrams (free body diagrams) [9], is based on the perspective of collaborative learning that is based on the psychological theory the sociocultural theory [7] and of cognitive development [10].

On the other hand, the force concept inventory (FCI) evaluates the degree of understanding of concepts in mechanical physics [11,12]. It is usually used as a pre-test/post-test (Hake's factor [13]) to determine the effectiveness of some teaching strategies used during students' conceptual learning. There are studies on the application of FCI in university students in the United States, Spain, and some Latin American countries but no studies are recorded in basic education students, which is where students are beginning to build their ideas of the physical concepts that will be used later in secondary and university education [3,14].

In accordance with the above, from the framework of the implementation of this strategy in the Physics subject in an educational institution of basic and secondary education, the hypothesis raised consists in whether the teaching of Newton's laws from the perspective of collaborative learning using force diagrams allowed a conceptual evolution of their learning. For this purpose, the FCI and Hake's factor were applied to know the students' conceptual level in relation to the concepts of mechanical physics.

2. Methodology

The research was done under the quantitative paradigm, with intentional sampling, because the target groups are formed through entry mechanisms outside of teachers' control. The design was quasi-experimental, in which the FCI was applied as a pre-test, to explore the preconceptions of Newtonian mechanics that students had, followed by a teaching intervention with the following activities: design, implementation and application of workshops based on real physical situations and where the elaboration of force diagrams is the central axis [7,10,11]. At the end of the intervention, the FCI was applied again as a post-test allowing to estimate the so-called Hake's learning gain [13].

2.1. Population and sample

The implementation of the proposal was developed in 40 of 10th grade students divided into two groups belonging to an educational institution located in Norte de Santander, Colombia. The ages of the students are between 14 years old and 16 years old. It should be noted that this is the first introductory physics course in their academic training. All the young people who participated in the implementation of the experience were informed orally about the nature and purposes of the experience; their response was of unanimous acceptance of wanting to participate actively in the process. In addition, they were informed of the commitment made with the institutional directives to deliver a copy of the project and its results, which will be available for any type of consultation.

2.2. Instruments

The FCI is a test designed by Hestenes, Wells, and Swackhammer that measures the understanding of the basic concepts of Newtonian mechanics, the didactic efficiency of the teaching-learning process of the latter and allows to detect the preconceptions that it has he evaluated on this subject [11,12]. The FCI is composed of 30 questions. The advantage of the FCI is that it allows to determine the level of knowledge of mechanics, to evaluate the didactic efficiency of the teaching-learning process, the degree of comprehension, to detect and to classify the conceptual errors incurred by the students and their preconceptions and their evolution in time [15,16]. The FCI was used in its Spanish version [3] with questions with 5 answer options, grouped into the following categories [11]: cinematics; Newton's first law (inertia); Newton's second law (force and acceleration); Newton's third law (action and reaction); principle of overlap; types of force. This instrument was used because it measures (in a certain sense) the ability of Newtonian thought [17]. A high score on the FCI does not indicate a unified knowledge of the concept of force, however, a low score indicates a lack of knowledge of basic Newtonian concepts.

Free-body diagram, also called force diagram [5] of a body or a group of bodies (or a part of a body) which is called a system and is represented in isolation from its environment, where all the external forces acting on it are shown, allowing the modelling of a body and its interactions with its surroundings. The modelling process allows to solve problems by simplifying the situations to be studied, therefore, it will be incomplete and inaccurate, but it will explain and predict the behavior of an object or system; the model will be valid if the analytical results are experimentally verified.

Finally, the Hake's factor or relative gain of conceptual learning indicates the average real gain of standardized conceptual learning [13]. It is used to determine the level of conceptual learning achievements in the implementation of a didactic strategy, that is, with the results of an evaluation (pre-test and post-test) the impact on the assimilation of type knowledge is determined conceptual. In the case of this proposal, the g factor allows to establish the changes achieved in the different dimensions of the FCI when implementing a didactic strategy, since the low, medium, and high levels of achievement in the g factor are related to the level of conceptual mastery of the phases of the FCI. For the calculation of the Hake's factor (g) the Equation (1) is used [13].

Also, the value of g can take values between 0 and 1, where 0 represents no learning, while 1 corresponds to the maximum possible learning. Establishing with the relative learning gain it is possible to classify three levels of achievement, high ($g > 0.7$), medium ($0.3 < g \leq 0.7$), and low ($0 \leq g \leq 0.3$).

$$g = \frac{FCI_{post}(\%) - FCI_{pre}(\%)}{100 - FCI_{pre}(\%)}. \quad (1)$$

2.3. Data collection procedure

The collection procedure is described below, which plays an important role in the validity and quality of the data. To analyze the initial state (preconceptions and diagnosis) that the students of mechanical physics had, the modified FCI in their numbering was applied as a pre-test (only 20 items were taken: 1-4, 7, 13-18, 21-29). The selected questions were directed to reach the conceptual levels of the students proposed in their area plan related to the concepts of force and proper Newtons laws. To perform the analysis, it was divided into fundamental aspects defined by groups of topics (see Table 1).

The classes were of a theoretical-experimental type where inclined planes, wooden blocks, laboratory pulleys, dynamometer, chronometers, pita ropes and power table were used. These activities were carried out in 20 hours of class attendance. For the didactic interventions related to the subject of cinematics (5 exercises) it was analyzed if the students have clear concepts of position, speed, and acceleration; as well as if they recognize these physical magnitudes as vector magnitudes.

The topic of Newton's Laws (10 exercises), it was identified if the students understand the approach of the law of inertia (movement does not imply presence of external force. We also analyzed the ideas they have about the cause of movement and its implications in speed and acceleration. Finally, the understanding of the interaction between two bodies was investigated, since it is common to consider a dominance principle in which "the strongest exerts the greatest force"; the strongest is usually the object with the greatest dimension, the most activity and/or the greatest amount of mass. This conflict leads to erroneous interpretations of this law.

Table 1. Classification of FCI questions by topic.

Topic	Grouping of questions
Cinematics	1-3, 5-15
Newton's first law	5, 7, 9, 10, 12-16, and 20
Newton's second law	1-3, 5-8, 11-15, 17, and 18
Newton's third law	4, 8-10, and 19
Force classes	1-3, 5-7, 11, and 20

The analysis of the subject each exercise consists of construction of force diagrams; decomposition of forces into rectangular components, if necessary; presentation of the equations of Newton's laws, and solution of the equations to find the physical variable requested in each statement, considering the data from each exercise. For the topic types of force (20 exercises), we explored the conception that they have of the term concept force; if they recognize the force of contact (friction and normal), the forces present by the action of the strings and the forces at a distance as the force of gravity. In addition, for the construction of force diagrams, the student considered the following protocol:

- Description of the physical situation.
- Simplified representation of the physical situation.
- Matrix representation of the interactions between systems.
- Representation with closed dotted line of the physical system(s) of interest.
- Detailed list of the forces acting on the physical system(s) of interest and assign a notation.
- Representation of the forces (force diagram) acting on the system(s) of interest.
- Representation of forces by simplifying the model to a particle (if possible).
- Once the situation of the exercise to be solved has been described and its simplified representation has been carried out, it should be done:
  - ❖ Choose reference frame.
  - ❖ Select a coordinate system that is anchored to the reference frame.
- Make the decomposition of forces (rectangular components) and proceed to apply Newton's laws of motion in each direction.
- Solve the equations.
- Verify the results by performing some usual technique for this purpose.

Finally, the FCI was applied again, and the so-called learning gain was estimated using the Hake's factor of each of the five aspects into which the study was divided.

3. Results and discussion

The presentation and analysis of the results was done considering the questions of the FCI (Table 2). The results are presented indicating the average percentage of success of the FCI questions in their pre-test and post-test applications and the calculation of the Hake's factor to numerically visualize the level of learning achieved (see Table 2). According to Hake's factor, the learning gain, for each of the subjects, is at the middle level, an index very similar to others.

Obtained in previous studies in the United States and Spain [18]. From the results of the table, it can also be highlighted that Newton's third law had the highest mean level of gain (0.63), which can be attributed to the analysis of physical situations with the use of force diagrams in association with the matrix listing of interactions. The above is due to the understanding of physical concepts rather than the mechanization of mathematical formulas and the proper construction of force diagrams.

In addition, for Newton's second law, it was the lowest level of profit (0.38). This is because some students failed to relate the acceleration to the resulting force; additionally, they presented deficiencies in the handling of the equations that in most of the cases were simultaneous. Here it is necessary to say that the time spent teaching this law was not enough. Some studies show that the concept with the greatest difficulty is Newton's second law [3,14].

Table 2. Average percentage of success in the pre-test and post-test and gain of the Hake's factor of the target group.

Topic	% Pre-test	% Post-test	Hake's factor	Level
Newton's first law	27.25	58.00	0.42	Medium
Newton's second law	31.92	57.85	0.38	Medium
Newton's third law	56.00	83.50	0.63	Medium
Force classes	32.15	58.71	0.39	Medium
Cinematics	39.45	68.21	0.47	Medium

Finally, between the application of the pre-test and the post-test a progress in the students' understanding and correct application of Newton's laws of motion is observed. The previous evidence shows that didactic intervention on the use of force diagrams (mental representations and their characterization), favored the significant learning of Newton's laws in basic education students, in comparison with previous experiences developed in similar contexts [19].

The realization of the force diagrams of the different physical situations allowed the student to evidence the understanding of the phenomena studied. In this order of ideas, the following advances in the learning of the concept of force and Newton's laws are highlighted.

- Construction of the interaction matrix, which contributed to differentiate the action-reaction pairs easily.
- Understanding sliding friction force.
- Correct interpretation of the friction coefficient.
- They adequately represent the external forces acting on the physical system.
- Improvements in mathematical procedures applied to the resolution of physical problems and interpretation of physical situations related to the environment.
- The feedback made it easier for the students to construct force diagrams.
- Resolution of more complex exercises.

Among the difficulties presented, the following stand out.

- The reference system or coordinates are not identified, and the physical variable requested is not found.
- Deficiencies in the handling of equations.
- Confusion of the concepts weight and mass, as well as in the nomenclature of forces.
- Difficulty in understanding physical situations such as underestimating friction, weightless pulleys and mass free ropes that can help simplify a problem.
- Incorrect decomposition of vectors into rectangular coordinates
- It is difficult in physical systems of three blocks to add masses and unite them into one.
- Difficulty to apply Newton's laws according to their cases.
- Difficulties in formalizing force diagrams.
- Although the reference systems and coordinates, the matrix, and the equations derived from the force schemes are well posed, they have mathematical difficulties in solving the problem.

Some of these difficulties in students coincide with those found in previous studies on physical concepts [20,21]; and especially concepts linked to the conception of Newton's laws [22-26]. Another of the limitations found was the decrease in the intensity of the weekly hours, due to the development of various academic activities at the institution.

According to the results obtained from the didactic implementation, and according to the instruments used, it is evident that the students have obtained gains with respect to conceptual learning, in the different components of the FCI [27]. Hake's factor does not seek to evaluate the student himself, but the teaching-learning process and the teacher's didactic strategies. In relation to the above, it is possible to establish that the use of force diagrams as a fundamental part in the teaching-learning of Newton's laws, allowed students to advance towards a high conceptual level, by improving their reasoning skills, developing an understanding of concepts, and solving more complex problems [28]. This, together with a better attitude of the students towards the study of physics.

4. Conclusion

After the analysis of the pre-test/post-test of the force concept inventory, it is shown that the didactic intervention that used the force diagrams as a fundamental element in the teaching-learning of Newton's laws allowed the systematic use of the interaction matrices of the forces of the physical systems, the experimentation of Newton's laws and their application to problematic situations achieving a significant learning in the students who were the objects of study. However, there is still much work to be done in this area since, although there are numerous proposals for didactic experiences of this type, their effectiveness must be evaluated with larger samples of students and those of different educational levels.

References

[1] Ministerio de Educación Nacional 1998 *Lineamientos Curriculares de Ciencias Naturales y Educación Ambiental* (Bogotá: Ministerio de Educación Nacional)

[2] Ministerio de Educación Nacional 2004 *Formar en Ciencias: ¡el Desafío!* (Bogotá: Ministerio de Educación Nacional)

[3] Budini N, Marino L, Giuliano M, Carreri R, Cámara C, Giorgi S 2019 *Revista de Enseñanza de la Física* **31** 107

[4] Machado C, Vieira P, Nascimento M A 2017 Combining two computational tools for teaching and learning physics *International Conferences New Perspectives in Science Education* (Florence: Pixel International)

[5] Tomara M, Tselfes V, Gouscos D 2017 *Themes in Science & Technology Education* **10(1)** 1

[6] Bello S 2004 *Educación Química* **15(3)** 210

[7] Vygotsky L 2009 *El Desarrollo de los Procesos Psicológicos Superiores* (Barcelona: Crítica)

[8] Ausubel D, Novak J D, Hanesian H 1983 *Psicología Educativa: Un Punto de Vista Cognoscitivo* (México: Trillas)

[9] Ocariz J 2000 Diagramas de cuerpo libre ¿Puede dibujarlos cualquiera? *Boletín del Departamento de Mecánica* (México: Univesidad Nacional Autónoma de México)

[10] Piaget J 1981 *Psicología y Pedagogía* (Barcelona: Ariel)

[11] Hestenes D, Wells M, Swackhamer G 1992 *The Physics Teacher* **30(3)** 141

[12] Hestenes D, Halloun I 1995 *The Physics Teacher* **33** 502

[13] Hake R 1998 *American Journal of Physics* **66(1)** 64

[14] Artamónova I, Mosquera J C, Mosquera J D 2017 *Revista Educación en Ingeniería* **12(23)** 56

[15] Huffman D, Heller P 1995 *The Physics Teacher* **33** 138

[16] Henderson C 2002 *The Physics Teacher* **40(9)** 542

[17] Morris G, Harshman N, Branum-Martín L, Mazur E, Mzoughi T 2012 *American Journal of Physics* **80(9)** 825

[18] Covián E, Celemín M 2008 *Enseñanza de las Ciencias* **26(1)** 23

[19] Prada Nuñez R, Hernández C A, Gamboa A A 2021 *Journal of Physics: Conference Series* **1981(1)** 012008:1

[20] Prada Nuñez R, Hernández C A, Gamboa A A 2021 *Journal of Physics: Conference Series* **1981(1)** 012007:1

[21] Elizondo M 2013 *Presencia Universitaria* **3(5)** 70

[22] Hestenes D, Wells M 1992 *The Physics Teacher* **30(3)** 159

[23] Hart G, Cottle P 1993 *The Physics Teacher* **31(8)** 470

[24] Roeder J 1998 *The Physics Teacher* **36(6)** 379

[25] Mora C, Herrera D 2013 *Latin-American Journal of Physics Education* **3(1)** 72

[26] Moreno J, Martínez Velásquez N 2017 *Amazônia: Revista de Educação em Ciências e Matemática* **13(26)** 82

[27] Hernández C A, Prada R, Gamboa A A 2021 *Journal of Physics: Conference Series* **2073(1)** 012003:1

[28] Leonard W J, Gerace W J, Dufresne R J 2002 *Enseñanza de las Ciencias* **20(3)** 387

AUTHOR INDEX